NAVAL
POSTGRADUATE
SCHOOL

MONTEREY, CALIFORNIA

THESIS

ANALYSIS OF HIGH ENERGY LASER WEAPON EMPLOYMENT FROM A NAVY SHIP

by

Ching Na Ang

September 2012

Thesis Advisor: Robert C. Harney
Co-Advisor: Douglas Nelson

THIS PAGE INTENTIONALLY LEFT BLANK

1. AGENCY USE ONLY *(Leave blank)*	2. REPORT DATE September 2012	3. REPORT TYPE AND DATES COVERED Master's Thesis	
4. TITLE AND SUBTITLE Analysis of High Energy Laser Weapon Employment from a Navy Ship		**5. FUNDING NUMBERS**	
6. AUTHOR(S) Ching Na ANG			
7. PERFORMING ORGANIZATION NAME(S) AND ADDRESS(ES) Naval Postgraduate School Monterey, CA 93943-5000		**8. PERFORMING ORGANIZATION REPORT NUMBER**	
9. SPONSORING /MONITORING AGENCY NAME(S) AND ADDRESS(ES) N/A		**10. SPONSORING/MONITORING AGENCY REPORT NUMBER**	

11. SUPPLEMENTARY NOTES The views expressed in this thesis are those of the author and do not reflect the official policy or position of the Department of Defense or the U.S. Government. IRB Protocol number _____N/A_____.

12a. DISTRIBUTION / AVAILABILITY STATEMENT Approved for public release; distribution is unlimited	12b. DISTRIBUTION CODE

13. ABSTRACT

This paper analyzes the employability of laser weapons on a Navy Littoral Combat Ship (LCS) class ship to counter small and fast boat threats. A general model of laser weapons is established to identify the attributes that characterize the laser weapon system. Quantitative values of each attribute are compared with current laser systems (that are under development) to identify potential laser types for employment on the Navy ship. In addition, plausible operational scenarios of suicide attacks by multiple (up to three) small and fast motor boats equipped with Improvised Explosive Devices (IED) against LCS-class ships are drawn up. These provide input parameters for computation of the required laser parameters to neutralize such threats. Based on the chosen laser technology and the calculated laser parameters, the requirements for employment of the laser weapon system on a Navy ship are determined.

14. SUBJECT TERMS Laser Weapon, Solid State Laser, Free Electron Laser, Fiber Laser, Littoral Combat Ship	15. NUMBER OF PAGES 87
	16. PRICE CODE

17. SECURITY CLASSIFICATION OF REPORT Unclassified	18. SECURITY CLASSIFICATION OF THIS PAGE Unclassified	19. SECURITY CLASSIFICATION OF ABSTRACT Unclassified	20. LIMITATION OF ABSTRACT UU

THIS PAGE INTENTIONALLY LEFT BLANK

ANALYSIS OF HIGH ENERGY LASER WEAPON EMPLOYMENT FROM A NAVY SHIP

Ching Na Ang
Civilian, Defence Science & Technology Agency, Singapore
B.S., National University of Singapore, 2005

Submitted in partial fulfillment of the
requirements for the degree of

MASTER OF SCIENCE IN SYSTEMS ENGINEERING

from the

**NAVAL POSTGRADUATE SCHOOL
September 2012**

Author: Ching Na Ang

Approved by: Robert C. Harney
        Thesis Advisor

        Douglas Nelson
        Co-Advisor

        Clifford Whitcomb
        Chair, Department of Systems Engineering

THIS PAGE INTENTIONALLY LEFT BLANK

ABSTRACT

This paper analyzes the employability of laser weapons on a Navy Littoral Combat Ship (LCS) class ship to counter small and fast boat threats. A general model of laser weapons is established to identify the attributes that characterize the laser weapon system. Quantitative values of each attribute are compared with current laser systems (that are under development) to identify potential laser types for employment on the Navy ship. In addition, plausible operational scenarios of suicide attacks by multiple (up to three) small and fast motor boats equipped with Improvised Explosive Devices (IED) against LCS-class ships are drawn up. These provide input parameters for computation of the required laser parameters to neutralize such threats. Based on the chosen laser technology and the calculated laser parameters, the requirements for employment of the laser weapon system on a Navy ship are determined.

THIS PAGE INTENTIONALLY LEFT BLANK

TABLE OF CONTENTS

LIST OF FIGURES

THIS PAGE INTENTIONALLY LEFT BLANK

LIST OF TABLES

THIS PAGE INTENTIONALLY LEFT BLANK

LIST OF ACRONYMS AND ABBREVIATIONS

ABL	Airborne Laser
AMM	Anti-Missile Missile
ANFO	Ammonium Nitrate and Fuel Oil
ATW	Atmospheric Transmission Window
BQ	Beam Quality
COIL	Chemical Oxygen Iodine Laser
CBC	Coherent Beam Combining
CIWS	Close-in Weapon System
DoD	Department of Defense
EO	Electro-Optical
FEL	Free Electron Laser
FELS	Free Electron Laser System
FL	Fiber Laser
GPS	Global Positioning System
IBC	Incoherent Beam Combining
IED	Improvised Explosive Device
JHPSSL	Joint High Power Solid State Laser
LASER	Light Amplification of Stimulated Emission of Radiation
LaWS	Laser Weapon System
LCS	Littoral Combat Ship
MASER	Microwave Amplification of Stimulated Emission of Radiation
MIRACL	Mid-Infrared Advanced Chemical Laser
MLD	Maritime Laser Demonstration

NA	Numerical Aperture
NAWC	Naval Air Warfare Center
Nd: YAG	Neodymium doped Yttrium Aluminum Garnet
NSWC	Naval Surface Warfare Center
OBT	Optical Beam Transport
ONR	Office of Naval Research
SBC	Spectral Beam Combining
SSL	Solid-State Laser
SSSL	Solid-State Slab Laser
TLS	Tactical Laser System
TNT	Trinitrotoluene
TRL	Technology Readiness Level
Yb:YAG	Ytterbium doped Yttrium Aluminum Garnet

EXECUTIVE SUMMARY

Three types of laser technologies, namely the Solid-State Slab Laser (SSSL), Free Electron Laser (FEL) and Fiber Laser (FL) are assessed for this thesis. The theory and technologies of high energy lasers are first introduced as a background to the topic. A general model of laser weapon system and characterization attributes is established. This model is being used to compare the existing laser technologies for being employed from the ship.

As the Navy is faced with rising asymmetrical threats, scenarios of suicide attacks from multiple (up to three) small and fast motor boats equipped with Improvised Explosive Devices (IED) against Littoral Combat Ship (LCS) class ships are developed and examined. Two sub-scenarios are considered, lasing directed towards the exposed IED that the target boat carries and lasing applied directly to the target's hull to hit the hidden IED. These scenarios are then used to determine the laser parameters (e.g., laser power and spot size) required to engage such types of adversaries.

The calculations and analysis determine that a 100 kW power laser would be adequate against non-hardened small and fast boats (e.g., fiber glass hull, exposed IEDs or out-board motors) carrying exposed an IED at a range up to 1 km. The size of such a laser system would be about that of a current Close-in Weapons System (CIWS) and requires the ship's electrical power of about 400 kW. This would be deemed adequate for employment on an LCS ship.

A boat target with an IED hidden within its aluminum hull is also explored. It would require a 1.6 MW power laser to burn through the aluminum hull to ignite the IED within a range of 1 km. The laser system would require an input electrical power of 6.5 MW, which exceeds the ship's electrical power at 3 MW. As a get around, lead acid batteries could be used to store the energy required for this laser system operation. With the weight and volume of the battery taken into consideration, the addition of the lead acid battery to support the operation of the 1.6 MW laser system is assessed to be reasonable. Also with the 1.6 MW laser system, the Navy ship would also be able to counter boats targets carrying exposed IED up to a range of 2 km in the absence of rain.

From the weight and dimension perspective, it would be possible to install the 1.6 MW laser system on the LCS-class ship. This may nevertheless compromise other onboard systems which have to be further studied.

Analysis has shown that it is not feasible to install the Free Electron Laser (FEL) on LCS-class ships mainly due to its size. In contrast, both the FL and SSSL should be able to match the ship's infrastructure capacities (weight, volume, electrical power, etc.). Looking at the current power achievement of the FL and SSSL, the SSSL appears to have higher potential of being employed on the LCS class ship in the near term as it has already reached the 100 kW power level. Nevertheless, an analysis of the trends of a single laser beam power shows that there is a potential for the FL to surpass the SSSL.

ACKNOWLEDGMENTS

The writing of thesis would not have been possible without the help of the following people that I would like to express gratitude towards.

In my endeavor, I was glad that I have a knowledgeable thesis advisor, Professor Robert C. Harney, who has been patient in providing guidance in his area of expertise, laser systems.

I would also like to thank Madam Mary Vizzini and Dr. Douglas Nelson for their help with this thesis.

Last but not least, I would like to take the opportunity to thank my husband, Kai Kok, for his unfaltering support and understanding throughout the past four years, especially while I was completing my master's degree at the Naval Postgraduate School.

THIS PAGE INTENTIONALLY LEFT BLANK

I. INTRODUCTION

A. RATIONALE

In recent years, there has been a rising trend of the use of asymmetric warfare techniques. Navy forces are facing small and fast targets (ships and supersonic missiles), some of them made at a very low cost. Hence, there is a need to take these targets out in a way that corresponds to the threat: with a low cost per shot, deep magazine (i.e., unlimited number of shots as long as input electrical power is available), high degree of precision and high speed.

The typical weapons available on a ship to engage such threats are general purpose machine guns, Close-In Weapon Systems (CIWS), naval artillery guns and anti-missile missiles (AMM). The cost and effectiveness of each of these weapons against such targets varies with different limitations and considerations.

General purpose machine guns are cheap but the range and accuracy are limited. The accuracy is limited largely due to the dispersion effects of guns. This weapon also depends heavily on the gunner's experience and skills.

A CIWS is effective up to a longer range of about 4 km with larger caliber and automatic targeting using radar system. Dispersion, although much lower than general purpose machine guns, is still large enough to require large number of rounds to guarantee a hit against a fast-moving target. When engaging multiple targets, it would run out of ammunition quickly.

Naval artillery guns are the next type of weapon available. This is a commonly available general purpose naval weapon, which could be operated as short-range anti-missile, anti-aircraft, anti-surface ship and ground support bombardment. However its effectiveness is averaged across its spectrum of operations. The main limitation of the naval artillery gun would be its accuracy. While the fire control radar could track its target, the radar is unable to determine the position of the gun projectiles with precision.

1

Hence, this makes it challenging to hit small and fast moving targets accurately. Proximity fuzing improves the probability of a hit, but does not completely compensate for modest accuracy.

Improvements to the accuracy of traditional shells (i.e., dumb rounds) have been made using guided artillery shells with its navigation aided by either laser or Global Positioning System (GPS). While the accuracy is improved, guided artillery shells have their own limitations. Laser guided shells require a laser designator to illuminate the target. However, laser designators used to beam the target are usually mounted on an aircraft that is not organic to the launch ship. Availability of the aircraft as well as ship-aircraft coordination would be challenging, especially against fast and small targets. For GPS guided artillery shells, their accuracy might not be good enough for small and fast targets.

The anti-missile missiles are the most effective against small and fast moving targets given their operating range, destructiveness, speed and accuracy. However, the cost is high and there is a limit on the number of missiles that can be carried onboard the ship.

Besides these existing classes of weapon, the next technology that has the potential to fill this operational requirement gap more effectively could be laser weapons.

The idea of delivering focused high energy across distance at speed of light is potentially a way to deal with such threats. As laser weapons have no need for ammunition, it is synonymous to having deep magazines. This allows the laser weapon to engage multiple targets at a low cost per target. Furthermore, with the high precision afforded by lasers, collateral damages are limited.

While the current, state-of-art laser weapons are still not yet capable of destroying targets instantaneously as portrayed in science fiction, laser technology has crossed the point where they can inflict enough damage to counter boats and drones with sustained lasing.

Solid-State Slab Lasers (SSSL) and Fiber Lasers (FL) have made transportable "tactical" applications possible on aircraft, ground vehicles and ships with their improved

2

efficiency and power levels. With the potential of high power, high beam quality, efficiency and frequency tunability, Free Electron Lasers (FEL) could offer defense against high maneuverability, sea-skimming supersonic missiles.

This paper studies three of the High Energy Laser (HEL) technologies (SSSL, FEL and FL) and analyzes how they could be effectively employed on existing Navy Littoral Combat Ship (LCS) class ships for practical operational usage.

B.   BACKGROUND: REVIEW OF LASERS AND LASER TECHNOLOGY

The fundamental science which enables the concept of the laser was the theory of stimulated emission published by Albert Einstein in 1917. The theory postulated the possibility of amplifying this form of emission. Practical realization to demonstrate this theory was not available until the later 1940s and early 1950s, where Charles Townes, Joseph Weber, Alexander Prokhorov and Nikolai G. Basov developed Microwave Amplification of Stimulated Emission of Radiation (MASER). The initial drive for this development was targeted at a microwave communication system. The MASER concept was subsequently applied on the optical frequency range of radiation, i.e., light. In 1960, Theodore Maiman demonstrated Light Amplification of Stimulated Emission of Radiation (LASER) using ruby as lasing medium, stimulated with high energy light flashes.

The ability to produce and direct a focused beam of energy to a fixed location at the speed of light opens up new applications in a wide variety of fields. One of these was in the area of military defense. There is plenty of science fiction portraying "death rays" which could instantly destroy targets, but in reality there exist many practical limitations and barriers to be overcome before the use of laser technology can be implemented. Hence, engineers and scientists have worked towards making this a reality.

There has been much research over the years conducted in laser technology, exploring various lasing media and pumping techniques that could be used to amplify stimulated emission. With development of different media and techniques, both engineering potential and constraints in practical application were revealed. Further work was needed to overcome these practical constraints to realize the potential of this technology.

The chemical laser, specifically the Chemical Oxygen Iodine Laser (COIL), was developed by the U.S. Air Force in 1977. The Airborne Laser (ABL) system which uses COIL technology was intended to demonstrate the capability of destroying airborne targets with gas lasers mounted on a large transport plane (McDermott, Pchelkin, Bernard, & Bousek, 1978). COIL works well at high altitude (i.e., a couple of kilometers) where the water-vapor content and atmospheric pressure are low. Consequently, the beam absorption is reduced. Contrary to high altitude, the water-vapor content and atmospheric pressure are high at sea level, resulting in a high laser beam absorption. This limits the performance of COIL at sea level.

In the early 1980s, the first megawatt-class laser emerged. The Mid-Infrared Advanced Chemical Laser (MIRACL) (FAS Space Policy Project, 1998) is a continuous wave, deuterium fluoride (DF) chemical laser that had been tested successfully against missiles at significant ranges. Despite its high laser power, MIRACL produces hazardous chemicals during emission, which could harm the crew as well as pollute the ocean.

As the megawatt-class chemical laser technology faced challenges with hazardous chemical discharge, another laser technology, known as the Free Electron Laser (FEL), was developed by John Madey at Stanford University in 1971 (National Research Council, 1994). In 1977, development reached greater heights following the successful operations of FEL at different wavelengths in the micrometer range. The demonstration of FEL tunability and design flexibility has since aroused great interest in FEL research.

During the 1990s, new technology developments in superconducting accelerators further improved the capabilities of the FEL technology. Success was achieved with the construction of a kilowatt-class continuous wave FEL proposed by the Navy at the Thomas Jefferson National Accelerator Facility (Williams, 2005).

While key limitations were identified and are yet to be resolved, interest in FEL persisted due to its ability to generate lasers at select frequency ranges which minimize

4

the effects of thermal blooming constraints. Thermal blooming is an atmospheric effect where the air absorbs a fraction of the radiation (heat) from the high energy laser beam. The air then expands resulting in reduced density in the laser path. The difference in density created between the air in the laser path and the surrounding area creates a lensing effect, which would defocus the beam. The amount of energy absorbed by air is a function of the laser wavelength; hence, the choice of frequency would greatly affect the effectiveness of laser weapon.

An alternative laser technology is the electrically pumped Solid-State Laser (SSL) that uses special solid materials as the gain medium. Tapping on the rise of semiconducting technologies, laser diodes are used as seed lasers for the SSL. The advantage of the SSL is its comparatively smaller size.

Another technology that is being developed is the Fiber Laser (FL). This leveraged on discoveries in the communication domain where light signals are amplified in optical cables in order to travel great distances. As the electro-optical repeaters used for signals amplification are not efficient, specially doped fiber-optic materials were developed. These materials allow amplification of light while signal, in the form of light, travels through the fiber. The improvement in pumping allows for better power efficiency of the FL. Other advantages of this technology are its compactness and commercial availability.

THIS PAGE INTENTIONALLY LEFT BLANK

II.   THEORY AND TECHNOLOGY OF THE HIGH ENERGY LASER

This chapter discusses the fundamentals of laser processes. While not all types of lasers (e.g., FEL) involve the transition of atoms between different electric energy states, this is a common method. Figure 1 shows the main components of a typical laser system.

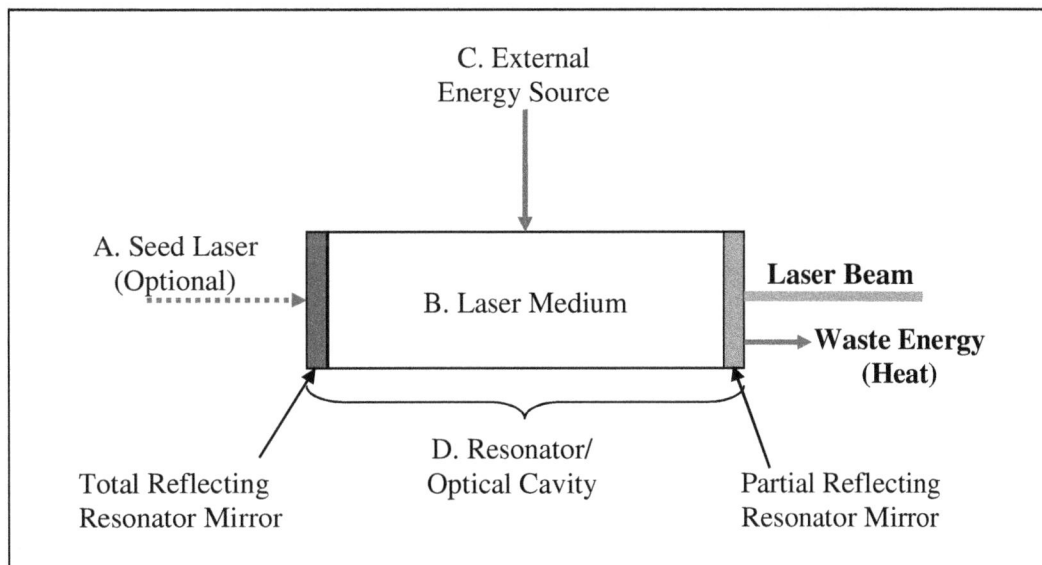

Figure 1.   Main components of a laser system

A.   SEED LASER (OPTIONAL)

The seed laser beam provides the initial source of light (seed light) into an amplifier or another laser for the lasing process to occur. Although a seed laser is an optional component in the laser system (an oscillator does not require a seed laser), it is often used in high energy lasers. The amplifier would boost the output from the seed laser while maintaining other characteristics of the seed laser beam, such as emission wavelength, polarization and short pulse width. A seed laser could be a diode or another laser source.

B.   LASER MEDIUM

The laser medium, which is also called gain or lasing medium, is the source of optical gain within a laser. This section describes the behavior of the atoms within the laser medium.

1.   Pumping Absorption and Emission

Normal matter is composed of atoms having discrete energy levels. An atom can move to an energy state by absorbing or emitting a photon. This is shown in Figure 2. During absorption, the atom takes up the energy from a photon and transitions to a higher energy level. The probability of this happening is proportional to the intensity of the light and the number of atoms that are currently at the ground state, N_1.

In contrast, an atom that travels to a lower energy state loses energy and could result in an emission of a photon. There are two types of emission, spontaneous emission and induced stimulated emission.

Spontaneous emission occurs when an atom is already at the excited state and undergoes a transition to a lower energy state randomly. This emits a randomly directed photon. Without a fixed phase or direction between emitted photons, spontaneous emission is not coherent.

Stimulated emission occurs when a photon strikes an excited atom, which induces it to emit another photon as it returns to the ground state. The photon emitted in such manner would have the same phase, frequency, polarization, and direction of travel as the incident photon. Hence, this stimulated emission would be coherent. The incident photon that struck the excited atom would not be affected. This coherence of photons generated from induced stimulated emission produces optical amplification, which is the basis of a laser system (Svelto, 1979).

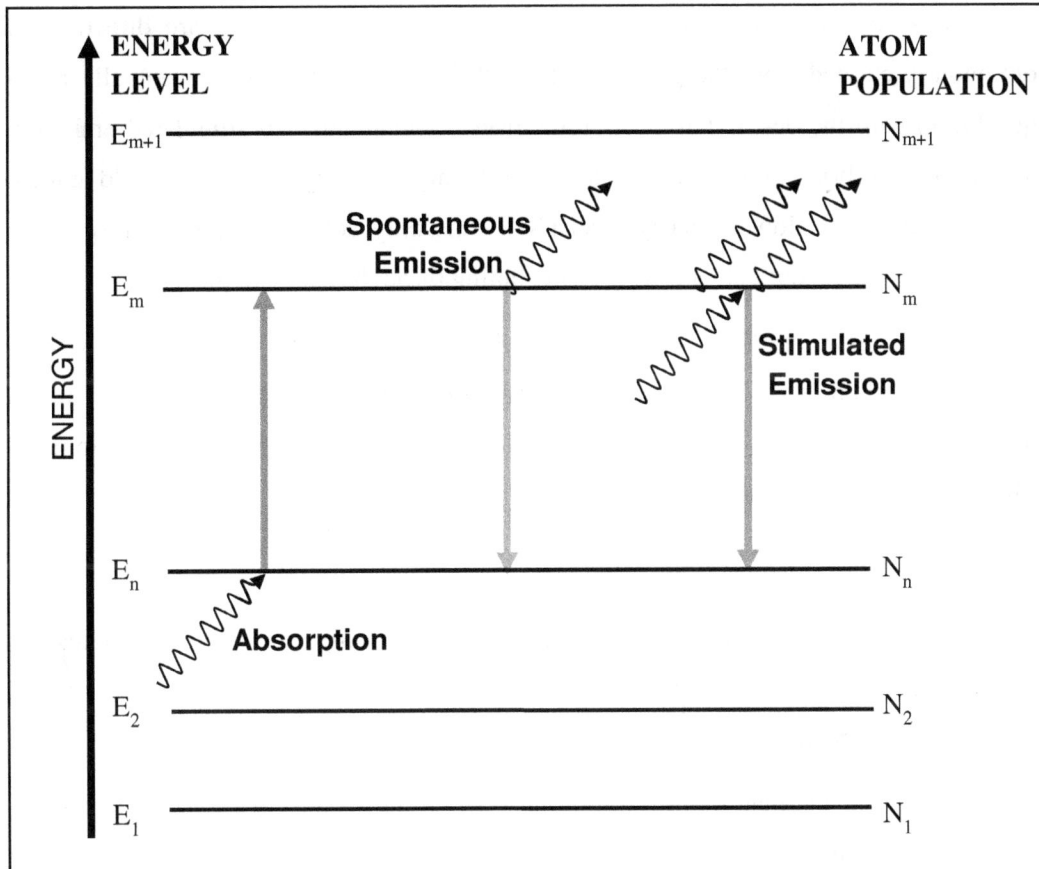

Figure 2.   Photon interaction processes in atoms (From Harney, 2012)

C.   EXTERNAL ENERGY SOURCE

As the name implies, the external energy source provides the energy to the laser medium to excite the atoms to initiate the lasing process.

1.   Population Inversion and Pumping

When the matter is in thermal equilibrium (or normal state), the number of atoms at the higher energy state (N_2) is small compared to that at the lower energy level (N_1). Population inversion is a condition where there is a greater amount of atoms at a higher energy state than the amount of atoms at some energy level below it. To achieve this, some form of activity is conducted to inject energy into atoms to raise them to a higher energy state.

However, with atoms of only two energy levels, atoms that are directly and continuously excited from the ground state to the excited state would eventually reach equilibrium with the de-excitation of atoms from spontaneous and stimulated emission. At thermal equilibrium, the number of atoms at the higher energy state (N_2) could at most equal the number in lower energy level (N_1). This only achieves optical transparency where there is no net optical gain from light shining through this medium.

To achieve non-equilibrium conditions where there would be optical amplification, there needs to be varying energy levels (i.e., more than three levels). As there are different rates of emission at each level, these varying levels allow atoms to build up at various intermediate states.

2. Three-level Laser

As an example, consider a three-level laser where there are three energy states, E_1, E_2, E_3, in increasing energy level, shown in Figure 3. At each energy state, there are N_1, N_2, N_3 atoms, respectively.

Initially at thermal equilibrium, almost all atoms would be at the ground state. Laser pumping transfers energy from an external source to the gain medium, exciting the atoms from ground state to Level 3 (known as the pump band). This is labeled as the pumping transition in Figure 3. While pumping is done most commonly by optical absorption (i.e., light energy), it can also be done by electrical discharge or chemical reactions.

By pumping the atoms continuously, a substantial number of excited atoms would be transited to Level 3 ($N_3 > 0$). To create a medium suitable for laser operation, these excited atoms need to quickly decay to Level 2. Theoretically, a photon may be released during this transition (spontaneous emission). However, in practice, this level of transition (from pump band Level 3 to Level 2) is usually radiation-less (i.e., without emission of photons) as the energy would only be converted to heat. This transition is denoted by R.

When an atom in Level 2 transitions to the ground state by spontaneous emission, it would generate a photon of certain frequency. This transition is known as the laser transition, L. If the lifetime of this transition is much longer than the lifetime of the radiation-less Level 3 to Level 2 transitions, a favorable lifetime ratio is established. In this condition, the atom population at E_3 would be essentially zero ($N_3 \approx 0$), as the excited atoms would accumulate in Level 2 ($N_2 > 0$). When more than half of the N atoms have accumulated in Level 2, population inversion ($N_2 > N_1$) is achieved between Level 1 and 2. This increases the number of stimulated emissions, which in turn leads to optical amplification. Frequency of the amplified laser would depend on the difference between energy Levels 2 and 1.

To ensure that more than half the population of atoms is excited from the ground state to obtain a population inversion, the laser medium must be very strongly pumped, which makes three-level lasers inefficient.

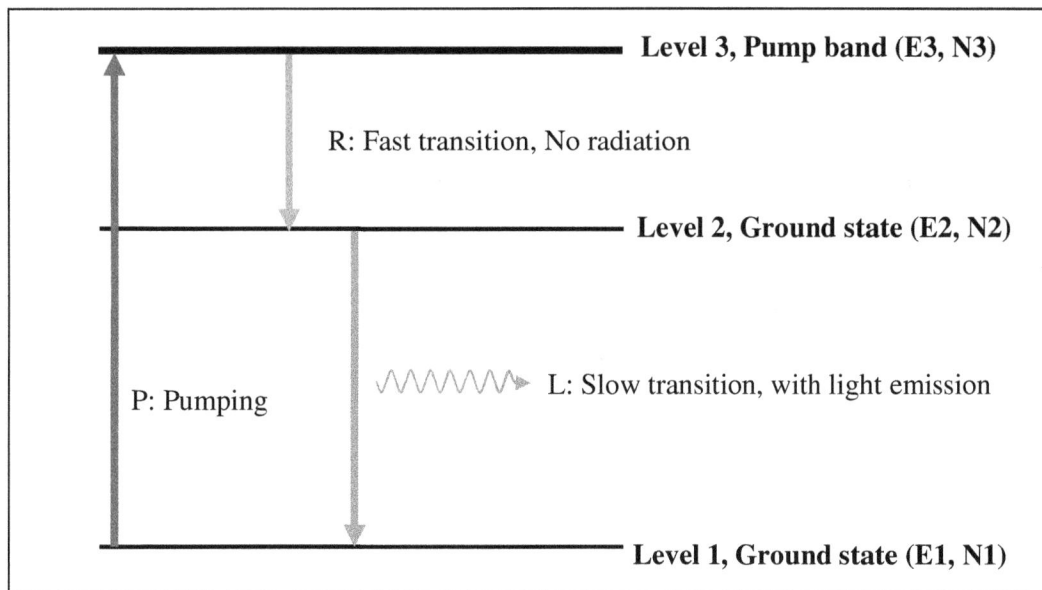

Figure 3.   Three-level Laser (From Mellish, 2005)

3. Four-level Laser

A four-level laser is used to make lasers more efficient in attaining population inversion. As illustrated in Figure 4, there are four energy levels, E_1, E_2, E_3, E_4, in increasing energy levels. Each has an atom population of N_1, N_2, N_3, N_4, respectively. Similarly, atoms are excited from the ground state to the pump band (Level 4 in this case) in the pump transition (P). The energy levels are specially designed to allow higher pumping efficiency. Level 4 atoms decay by a fast, non-radiative transition (Ra) to Level 3. As the lifetime of the laser transition, L, is long compared to that of Ra, many atoms would accumulate in Level 3 (upper laser level) which may then transit to Level 2 (lower laser level) by spontaneous or stimulated emission. Level 2 also has a fast and radiation-less decay, Rb, to the ground state.

Similar to the three-level laser, the atom population at the pump band (E_4) would be essentially zero ($N_3 \approx 0$), as the excited atoms would accumulate in Level 3 ($N_3 > 0$).

In a four-level system, atoms in the lower laser level, E_2, get de-excited quickly such that the numbers of atoms in that level is essentially zero ($N_2 \approx 0$). As a result, only a small number of atoms in Level 3 ($N_3 > 0$ and $N_3 > N_2$) would form a population inversion with respect to Level 2, which is required for optical amplification to take place. Frequency of the amplified laser would depend on the difference between energy Levels 3 and 2.

12

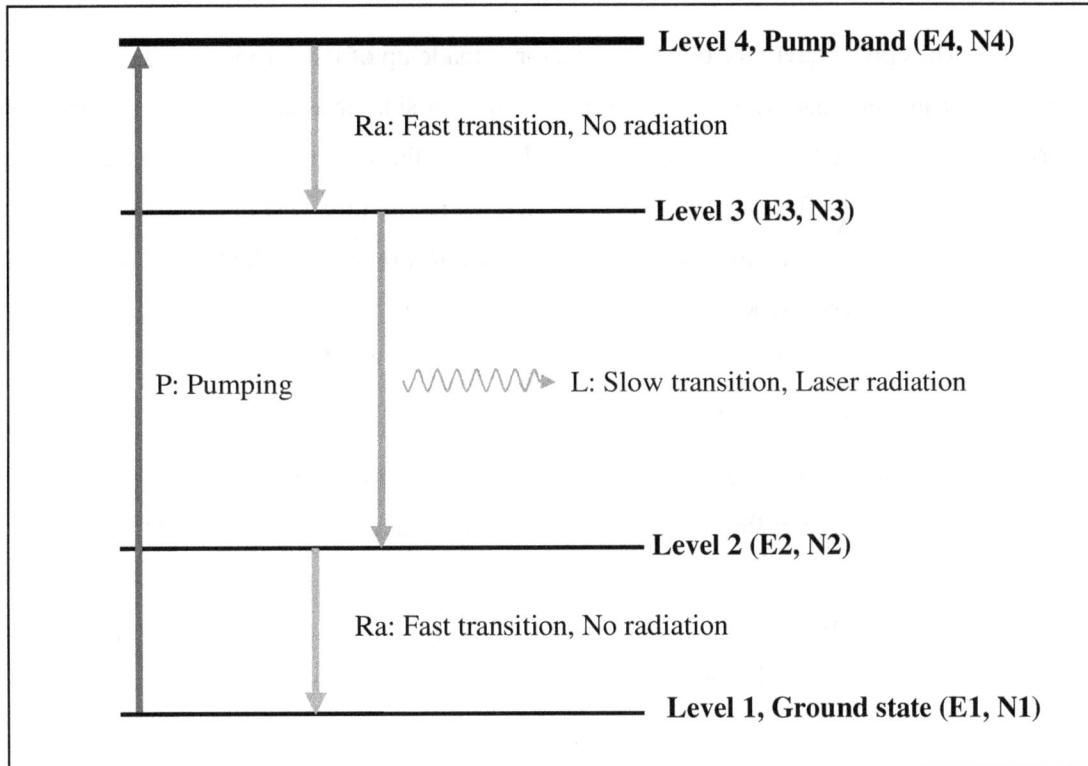

Figure 4.   Four-level Laser (From Mellish, 2005)

The three-level laser requires more than half of its atoms to be excited to the pump band (N_3) to achieve population inversion ($N_2 > N_1$). However, in the case of the four-level laser, it only requires a few atoms in the pump band (N_4) to obtain a population inversion, ($N_3 > N_2$) since N_2 is always nearly zero. Hence a four-level laser has better pumping efficiency in attaining population inversion. Most practical lasers are four-level and higher levels are not uncommon. The main purpose of using more levels is to allow optical pumping of the medium at a wide range of wavelengths.

Regardless of the number of levels, the energy of the pumping transition has to be larger than the laser transition's to form a population inversion. For optical pumping, the frequency of the incident light used must be greater than that of the resulting laser light as higher frequency light has higher energy.

13

D.   OPTICAL CAVITY

The optical cavity or optical resonator is made up of two or more (e.g., ring laser) resonator mirrors that are specially arranged on each side of a gain medium to trap light (photons). At specific separation distances between the mirrors, there is no destructive interference from reflected wave fronts. Hence, when the direction of the photon emitted from this medium is parallel to the optical axis, it would be reflected by the mirrors causing it to traverse back and forth within this cavity across the lasing medium, forming standing waves at a fixed resonant frequency. Photons would only be able to build up their energy without interference.

The motion of reflected photons induces stimulated emissions that travel in the same direction towards the fully reflective mirror. When the photon hits this mirror, it reflects again and travels back towards the partially reflective mirror. Again, as it travels through the gain medium, more photons would be generated. This process amplifies the light (resulting in a gain in light intensity) within the optical cavity.

With continuous pumping of the medium, a constant supply of energetic atoms would be available for continued stimulated emissions. While some of the photons are reflected at the partially reflective mirror, some leak, which contributes to loss of light from the optical cavity. As stimulated emission continues in the optical cavity, it increases the intensity within the cavity.

A higher intensity also increases the loss through the partially reflective mirror. A steady state would eventually be reached when total loss of light from the partially reflective mirror equals the gain from the optical cavity. The constant loss of high intensity light from the partially reflective mirror constitutes the laser energy beam that would be directed at the target.

III. DIFFERENT TYPES OF LASER SYSTEMS

This chapter looks at three types of high energy lasers that are used or could potentially be used for laser weapon systems. They are the Solid-State Slab Laser, Free Electron Laser and Fiber Laser. All lasers work based on the fundamental theory and general system model made up of a lasing medium, pump source and optical cavity. The key difference lies in the technology used to achieve the functional purposes of each sub-component and the sub-component's impact on the overall system design.

A. SOLID STATE SLAB LASER

As the name implies, the Solid-State Laser relies on solid as the lasing medium. For the purpose of this thesis, only the Solid-State Slab Laser is examined. The SSSL works as described in the fundamental theory of general lasers, where there is a laser medium, pump source and optical cavity.

1. Slab Lasing Medium

Slab lasers are one class of high power solid-state bulk lasers. As the name implies, the gain medium (laser crystal) has the form of a slab. Generally, a laser slab is thin on one dimension compared to the other two dimensions (each slab or block has three dimensions). Some advantages of slab lasers over rod lasers include better beam quality, lower stress induced birefringence and higher optical-to-optical efficiencies. Birefringence, also known as double refraction, occurs when a light ray is split into two rays as it passes through anisotropic materials, such as crystals. Slab lasers will be elaborated in greater detail in the subsequent sections.

2. Pump Source

The two main pump sources for Solid-State Lasers are diode pumps and flash lamps. In general, flashlamps are cheap but inefficient, whereas diode pumps are efficient but expensive. For the purpose of this thesis where it focuses on high energy lasers, only diode pumps will be discussed.

To achieve high average power, neodymium and ytterbium doped yttrium aluminium garnet (Nd:YAG and Yb:YAG) are frequently used as the gain medium. It has been shown that these two materials have pump absorption efficiencies, η, of approximately 80% (Bass, 2005). η is the ratio of light absorbed by the medium to the light radiated from the pump source (i.e., laser diode) at the pumping frequency. The advantages of diode-pumped SSLs are their compactness and efficiency.

For diode pumping of slab lasers, edge-pumping (also known as transverse-pumping) geometry is more suitable vis-à-vis the face pumping geometry. Both geometries are provided in Figure 5. As the high power laser diodes have a high degree of spatial coherence, they could inject all pump light into the thinner edge. Since it is not required for the other two dimensions (or faces of slab) to be transparent, a larger range of cooling mechanisms are applicable. This delinks the cooling and optical pumping interfaces, which allows a less complex laser design, as well as creates a long path for absorption along the wider dimension (Rutherford, 2001). With the long path, a very thin slab can be used without affecting the pump absorption efficiency. Other advantages of such geometry include scalability and uniform conductive cooling.

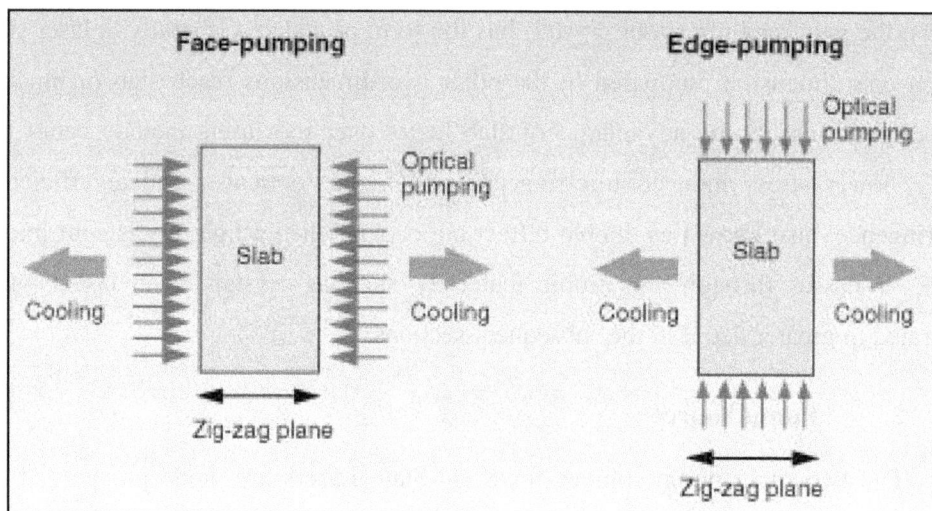

Figure 5. Face and Edge-pumping (From Tyler, Korczynski, & Sumantri, 2000)

Due to the pronounced asymmetry of both geometry and thermal lasing, power extraction from slab lasers is difficult. One way to reduce the strong thermal lensing in one direction is to let the laser beam make a zig-zag path through the gain medium, which is known as the zig-zag slab geometry, as illustrated in Figure 6. By doing so, the effects of the strong thermal lasing in the thinner dimension would be averaged out. Another way for power extraction is the use of an unstable laser resonator (Paschotta, 2008).

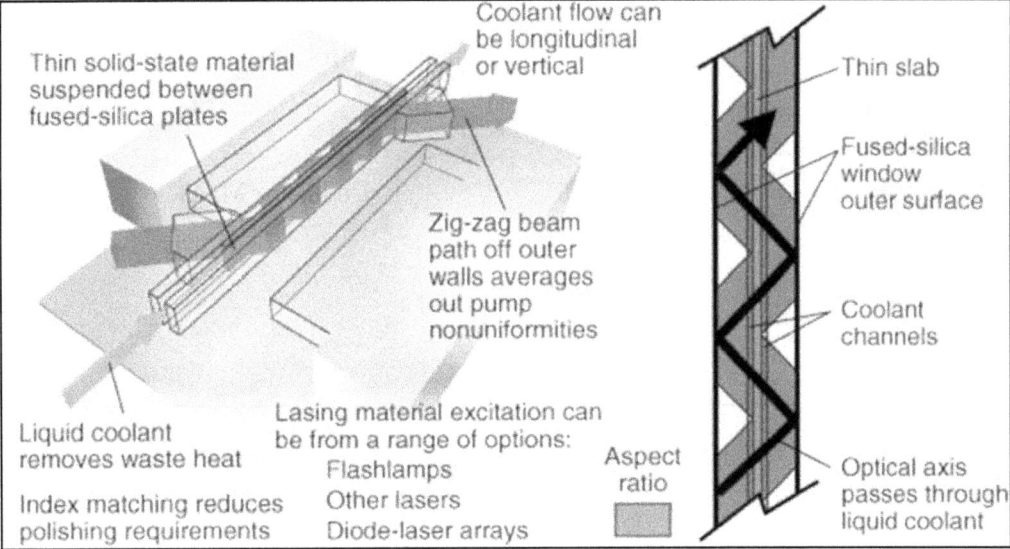

Figure 6.   Zig-Zag Slab Geometry (From Hecht, 2007)

3.   Advancement of High Energy Solid-State Lasers

In 2009, Northrop Grumman demonstrated its scalable building block approach for compact electric laser weapons. With the combining of several 15 kW laser building blocks, a 100 kW light ray was created by an electric laser under the U.S. military's Joint High Power Solid State Laser (JHPSSL) program. The achievement included turn-on time of less than one second and continuous operating time of five minutes (Northrop Grumman). This achievement could be a proof of principle for weapons-grade power levels for high energy lasers. While the SSSL has potential for high energy laser weapons, it has a low wall-plug efficiency.

B.   FREE ELECTRON LASER

Unlike the other types of lasers, the FEL does not have a medium which contains bound electrons. Instead, it uses a high-energy electron beam as an amplifying medium. A simplified diagram of the process is illustrated in Figure 7 and the subsystems are elaborated as follows.

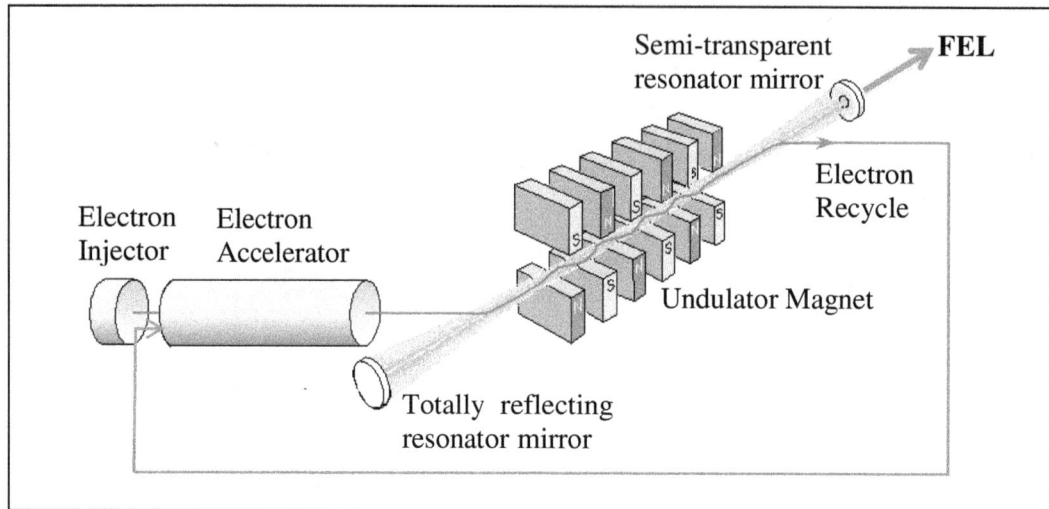

Figure 7.   Basic parts of a Free Electron Laser (After Willingale, 2007)

1.   Electron Injector

The electron injector (or electron gun) is basically a photocathode that extracts electrons from a metal surface via photoelectric effect.

2.   Electron Accelerator

Emitted electrons from the injector are accelerated to relativistic velocities (nearly the speed of light) by an electron accelerator.

3.   Undulator

The undulator (or wiggler) is made up of a series of magnets with alternating north and south poles. When the accelerated electrons pass through the alternating magnetic field, they would bend in a sinusoidal fashion according to the Lorentz Force

Law. As a result, photons are emitted along the undulator axis. The undulator allows frequency tuning by varying the undulator's field strength (amplitude and period) or the timing of the electromagnetic path.

4. Optical Cavity

As the photons leave the undulator, they traverse to the semi-transparent resonator mirror that allows some photons to pass through, while reflecting the majority of the photons up the beam line. At the other end of the beam line, there is a totally reflective resonator mirror. These two pieces of resonator mirrors form the optical cavity. As reflected photons interfere constructively with the other photons along the undulator repeatedly, it leads to a growth in intensity and coherence of the emitted light.

The electrons interact with their own spontaneous emission, which in turn amplifies the EM field. This process iterates and builds up (saturates) until a steady state FEL beam is formed.

5. Electron Recycle and Electron Dump

While some FEL designs use electron dumping to absorb the high-energy electrons that cannot be recovered during the lasing process, there are some designs that recycle them prior to dumping them. After passing through the undulator, the unrecovered electrons still retain much of their kinetic energy. Hence, by sending these electrons through the accelerator, they can be induced to give up (recycle) their kinetic energy to those newly injected electrons from the injector. Recycling electrons reduces thermal waste which would improve the wall-plug efficiency of the laser system. Additionally, it also decreases the energy of the dumped electrons, which reduces radiation.

6. Advancement of High Energy Free Electron Lasers

By varying the input electron energy or the undulator's field strength (amplitude and period), the FEL is tunable across a wide spectrum of wavelengths, from microwave through terahertz radiation and infrared to the visible spectrum to ultraviolet to soft X-rays (Horn, 2009). The ability to select a specific wavelength made it possible to tune the beam to local atmospheric conditions to maximize its performance.

In 2006, Jefferson Laboratory achieved a FEL power of 14 kW at 1.6 μm (Defense Science Board, 2007), a wavelength suitable for maritime propagation. Following that, the Navy proposed to develop a 100 kW FEL system, as a stepping stone to scale to the megawatt power level in the infrared FEL technology. This serves as a study for FEL integration on future Navy ships to provide ship defense (Office of Naval Research, 2008). Another advantage of the FEL is its high beam quality.

While the FEL could be scaled to megawatt power level using a superconducting electron gun and accelerator, it requires high amount of electrical input power and cooling due to its low wall-plug efficiency (Allagaier, 2003). As a result, the cost of the FEL would be high. The huge size of the FEL would also pose a constraint to platform integration.

C. FIBER OPTICS LASER

The Fiber Laser works similarly to the SSL. In fact, some consider the FL as a SSL. Figure 8 illustrates the main components of a FL.

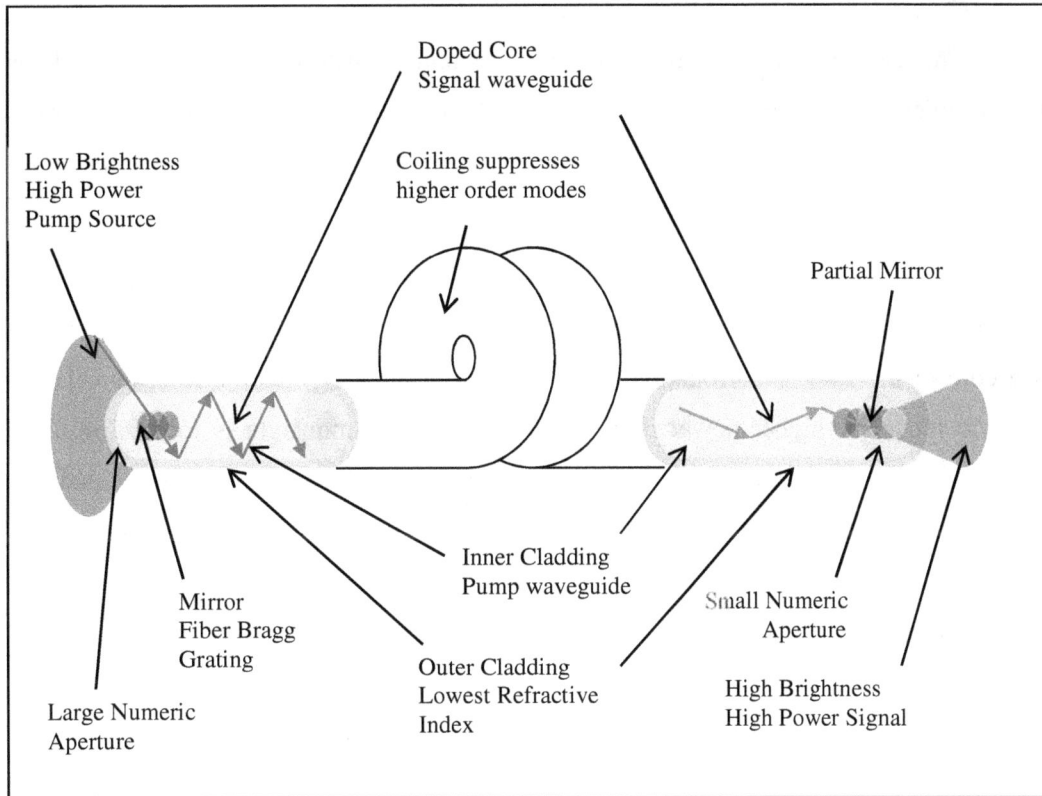

Figure 8.   Basic parts of a Fiber Optics Laser (After Motes & Berdine, 2009)

1.   Optical Fiber Medium

For fiber optics lasers, the active gain medium is an optical fiber doped with rare-earth elements such as erbium, ytterbium and neodymium. The gain medium forms the core of the fiber.

2.   Pump Source

The pump source for a FL could be semiconductor diode lasers or other fiber lasers. Diode lasers can be stacked to obtain a higher power to pump a single fiber laser with large amounts of power.

3. Double-Clad Fiber

The usual optical fiber has two layers of optical materials where pump light is directly injected into the single-mode core. As the core has a small Numerical Aperture (NA), the pumping of the divergent output beam from the pump source becomes challenging until dual-clad technology arises.

A double-clad fiber has three layers (core, inner cladding and outer cladding), as shown in Figure 8. The three layers are made of materials with different refractive indices. The core of the fiber is surrounded by two layers of cladding. As the fiber core is too small to focus the higher power diode laser into it, the pump light is focused into the much larger inner cladding around the core. As light travels through the inner cladding via Total Internal Reflection (TIR), it would pass through the signal core that is embedded within the inner cladding. The outer cladding keeps the pump light contained within. With this arrangement, it creates a much higher power beam within the fiber to pump the atoms in the core.

4. Laser Cavity

Contrary to using mirrors in the cases of solid-state and free electron lasers, Bragg gratings are added to create the laser cavity. A Bragg grating is a section of glass that has stripes in it to vary the refractive index periodically along the length of the fiber core. When a broad spectrum light is shined into one end of fiber containing a fiber Bragg grating, the light with wavelength that matches the Bragg grating wavelength would be reflected back to the input end, while the rest of the light would be transmitted through the Bragg grating. The Bragg grating acts like a mirror.

5. Advancement of High Energy Fiber Lasers

The current state of art for the single power fiber laser is 6 kW (Motes & Berdine, 2009). As the beam is not narrow-band with single-polarized output, it could not be coherently combined with other fibers to produce good beam quality at high power at this time. While high power commercial fiber lasers do not have beam quality good enough for military applications yet, this could be achieved in the near future as the development of fiber lasers are not only of interest to the military but also to commercial industries.

As the laser beam is generated inside the small core of the fiber, the delivery of the beam does not require complicated or sensitive optics. This makes fiber optic lasers very stable and easy to use. With the beam confined within the small core of the fiber, the laser beam produced is of high quality (i.e., it can be focused on very small dot). A fiber optics laser is also able to convert a larger proportion of power delivered by the pump source into the laser beam, (i.e., better wall plug efficiency) as compared to the other laser types. As the fibers are generally long, heat generated is distributed over the length of the fiber, which helps in cooling (Hunter & Leong, 1996). Although fiber lasers have many advantages, they have a disadvantage for power scaling. Due to the confinement of the optical energy within the waveguide, there are very high field intensities that could cause nonlinear optical effects in the fiber. As a result, high power, continuous wave fiber laser performance could be degraded (Stephen, 2008).

D. PROS AND CONS OF DIFFERENT LASERS IN GENERAL

While the three types of lasers use the same physics principles, there are several properties that make them more useful or constraining than others. Table 1 summarizes the advantages and limitations of each of the three lasers as described in earlier sections.

	SSSL	FEL	FL
Advantages	- High beam quality (single slab) - Scalable power via beam combining	- Tunable wavelength - Can be made to go up to very high powers - High beam quality	- High beam quality (single fiber) - Efficient and easier to cool - Development fuelled by commercial applications - Scalable power via beam combining
Limitations	- Low wall plug efficiency	- Large system size - Very low wall plug efficiency - High cost	- Undesired nonlinear optical effects at high energy could degrade performance - Low wall-plug efficiency

Table 1.   General High Energy Lasers

IV. ANALYSIS OF LASER WEAPON TECHNOLOGY

With the introduction of the three types of laser technology in Chapter III, this chapter discusses their applications in laser weapons, with reference to the current developmental laser weapon systems for the U.S. Navy. These four systems are the Maritime Laser Demonstration (MLD), Free Electron Laser System (FELS), Laser Weapon System (LaWS) and Tactical Laser System (TLS). The purpose of the analysis is to determine their suitability for practical employment on existing naval platforms.

This involves the preliminary step to establish a general model of laser systems and identification of various key attributes which characterizes the system. This serves to provide quantitative comparisons between laser systems and a common understanding for qualitative discussions.

Subsequently, the power requirements for engagement of different targets would be highlighted. This is important to determine the applicability of the system to the operational needs of the platform.

Lastly, as each technology progresses at different rates due to their unique strength and weaknesses, it is important to consider time frames when selecting different laser technologies for employment. The plan would likely be in the form of a progressive road map with attempts at early adoption for supporting applications with room for capability enhancements and upgrades to expand its responsibility as technology matures.

A. KEY ATTRIBUTES OF LASER WEAPON SYSTEMS

Following the lasing process as described in Chapter II, a laser beam is produced. The following attributes are used to determine the effectiveness of the laser beam. They would also form the model to assess a laser beam.

1. Beam Power

Beam power is one characteristic of the laser beam that would determine the degree of destruction that the laser can inflict on the target. Measured in watts (W, kW and MW), the laser beam power refers to the optical power output of the laser beam.

Before discussing the laser technologies, it is pivotal to identify the laser power level required to counter different classes of targets. Table 2 shows the approximate laser power levels for different targets from the perspective of Navy, research organizations and industry.

As the target gets more capable in terms of speed and maneuverability, a laser with higher beam power is required to counter it. Current technology, which would be discussed in later sections, may be adequate to counter small boats. Nevertheless, to attack missiles, megawatt level power would be necessary.

Source	Beam power measured in kilowatts (kW) or megawatts (MW)				
	~ 10kW	Tens of kW	~ 100kW	Hundreds of kW	MW
One Navy briefing (2010)		UAVs			
			Small boats		
				Missiles (starting at 500kW)	
Another Navy briefing (2010)		Short-range operations against UAVs, RAM, MANPADS (50kW-100kW; low BQ)		Extended-range operations against UAVs, RAM, MANPADS, ASCMs flying a crossing path (>100kW; BQ of ~2)	Operations against supersonic, highly manoeuvrable ASCMs, transonic air-to-surface missiles, and ballistic missiles (>1MW)
Industry briefing (2010)		UAVs and small boats (50kW)	RAM (100+ kW), subsonic ASCMs (300kW), manned aircraft (500kW)		Supersonic ASCMs and ballistic missiles
Defense Science Board (DSB) report (2007)		Surface threats at 1-2km		Ground-based air and missile defense, and countering rockets, artillery, and mortars, at 5-10km	"Battle group defense" at 5-20 km (1-3 MW)
Northrop Grumman research paper (2005)	Soft UAVs at short range	Aircraft and cruise missiles at short range	Soft UAVs at long range	Aircraft and cruise missiles at long range, and artillery rockets (lower hundreds of kW) Artillery shells and terminal defense against very short range ballistic missiles (higher 100s of kW)	

Table 2.   Approximate Laser Power Levels required for different targets (From O'Rourke, June 2012)

2. Beam Quality

Beam quality (BQ) is a measure of how well the beam can be focused on a point. Both Beam Parameter Product (BPP_A) and M^2 are measures of BQ.

BPP_A is defined as the product of full (far-field) beam divergence angle, θ and the diameter of the beam at the waist, W. The beam waist, also known as the beam focus of a laser beam, is the location along the propagation direction where the beam radius is the minimum.

$$BPP_A = \theta W \qquad \text{Equation 4-1}$$

M^2 is the ratio of BPP_A and the beam parameter product of a diffraction-limited Gaussian beam, BPP_G, with the same wavelength. A diffraction-limited Gaussian beam is an ideal beam, where the beam is best focused for the given wavelength.

$$M^2 = BPP_A / BPP_G \qquad \text{Equation 4-2}$$

For this thesis, only M^2 would be used as an indicator for BQ. The best achievable beam quality happens when BQ or M^2 equals to 1. Having a BQ of 2 is akin to having a laser's light spot that is twice as large in diameter as a same laser with a BQ of 1 at a given range. This also means that the intensity of the beam (which is affected by the area of the spot size) is decreased by a factor of 4.

3. Wavelength

While most of the electromagnetic radiation from the laser beam is absorbed by the atmospheric gases and particles, some of the radiation is transmitted through the atmosphere. These regions that are not or are less affected by atmospheric absorption are known as the Atmospheric Transmission Windows (ATW) shown in Figure 9. To minimize the effect of absorption, which would affect the beam power and quality, the wavelength of the laser beam is selected to match the atmospheric transmission windows. Other than operating within the transmission window, the operating wavelength should be selected such that the incident or reflected energy should minimize or would not cause eye hazards to humans. Table 3 provides a summary of the effects of laser wavelengths on the human eye.

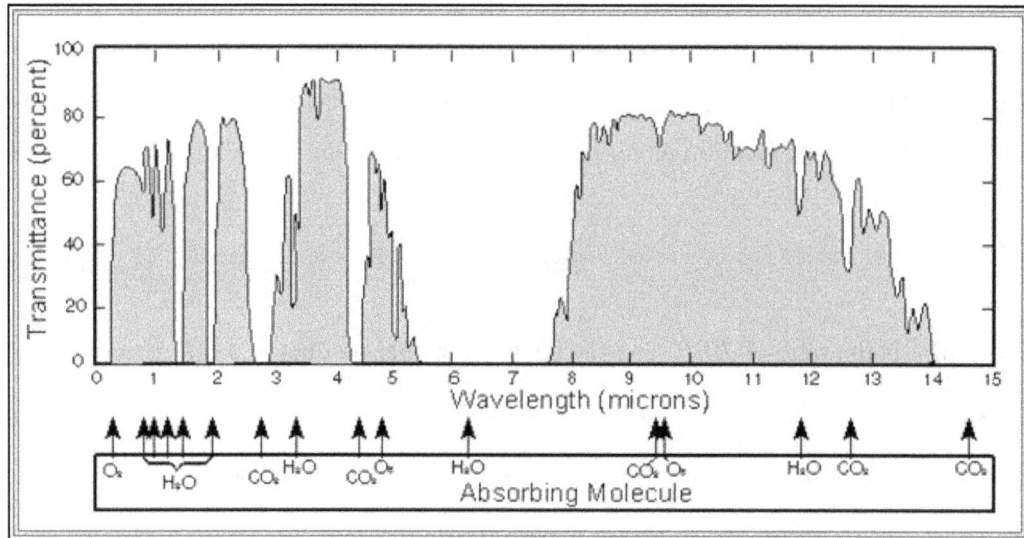

Figure 9.   Atmospheric Transmission Window (NITEHOG Systems LLC, 2012)

Wavelength	Area of Damage	Pathological Effect
180 - 315 nm (Ultraviolet UV-B, UVC)	CORNEA; Deep-ultraviolet light causes accumulating damage, even at very low power	Photokeratitis; Inflammation of the cornea, similar to sunburn
315 - 400 nm (Ultraviolet UV-A)	CORNEA and LENS	Photochemical cataract; Clouding of the lens
400 - 780 nm (Visible)	RETINA; Visible light is focused on the retina	Photochemical damage; Damage to retina and retinal burns
0.78 – 1.4 µm (Near Infrared)	**RETINA; Near IR light is not absorbed by iris and is focused on the retina**	**Thermal damage to cataract and retinal burns**
1.4 – 3.0 µm (Infrared)	**CORNEA and LENS; IR light is absorbed by transparent parts of eye before reaching the retina**	**Aqueous flare; Protein in aqueous humor, cataract, corneal burn**
3000 – 10000 nm (Far Infrared)	CORNEA	Corneal burn

Table 3.   Effects of Laser Wavelength on the eye (From UCLA Laser Safety Lite, 2009)

4. Laser Wall-Plug Efficiency

Laser wall-plug efficiency is how well the laser system converts input electrical power into laser beam power output. It is indicated by percentage (%). The loss in power is converted to heat energy. Depending on the amount of heat generated, an appropriate cooling system would be required to prevent overheating of the system, which would affect the performance of the system.

5. Weight and Dimension

The Navy LCS does not offer the luxury of space. Hence, it is important that the weight, size and power consumption of the laser system fit the existing infrastructure of the ship. Besides the optical subsystems, the laser system also requires supporting equipment such as an electrical power system, a cooling system and radiation shielding to protect the equipment and/or personnel from hazardous radiation.

B. CURRENT NAVAL LASER SYSTEMS UNDER DEVELOPMENT

With many advantages of laser weapons as described in Chapter I, the U.S. military has been putting much emphasis on developing high energy lasers for weaponry applications. This section studies four laser systems which are currently under development by Department of Defense (DoD) for naval applications. The information on the four systems is summarized in Table 4.

1. Maritime Laser Demonstration (MLD) – SSSL
(O'Rourke, 2012; Defense Update, 2009)

The MLD is a prototype laser weapon system that is under the DoD's Joint High Power SSL (JHPSSL) program. Tapping on the other development of slab lasers under JHPSSL, e.g., Firestrike, Northrop Grumman made a coherent combination of seven slab laser beamlets of 15 kW each to produce MLD's laser beam of 105 kW. Coherent beam combination occurs when all of the beam array elements operate with the same spectrum and the relative phases of the elements are controlled such that there is constructive interference (Fan, 2005). The system has been tested in sea environment.

Figure 10.   Maritime Laser Demonstration (From O'Rourke, June 2012)

2.   Free Electron Laser System (FELS)
(Office of Naval Research, 2008)

Due to the large size of the FEL, its development has only been done for the Navy, which carries bigger platforms. While a 14.7 kW FEL has been produced and has been assessed that the power can be scaled to megawatts without other technology breakthroughs, the system is still under laboratory testing.

Figure 11.   Free Electron Laser (Jefferson Lab News, 2006)

3.   Laser Weapon System (LaWS) – FL
(O'Rourke, 2012)

As a possible add-on to the Close-in Weapon System mount, LaWS is designed to disable Electro-Optical (EO) sensors, and counter Unmanned Aerial Vehicles (UAV) and EO guided missiles. Today, it has achieved a beam power of 33 kW by incoherently combining six fiber laser beamlets in an over-the-water test environment. Incoherent combining basically means that the beams are not in the same phase. There are two types of incoherent beam combining concepts. The first is spatial (or side-by-side) beam combining in which the array elements may (or may not) operate at the same wavelength, but nothing is done to try to control the relative spectra or phases of the elements (Fan, 2005). For spatial combining, power will be increased while the beam quality will be worse. The second one is spectral beam combining which occurs when the beam array elements operate at different wavelengths and then a dispersive optical system is used to overlap the beams from the elements in the near and far fields (Fan, 2005). For spectral combining, the power will increase while the beam quality remains the same.

Figure 12.   Laser Weapon System (From O'Rourke, June 2012)

4.   Tactical Laser System (TLS) – FL
(Defense Update, 2011)

Having a similar concept as the LaWS, the TLS is envisioned to be added onto the Mk38 25 mm machine gun. The TLS is designed to have a single phase (without beam combining) laser beam power of 10 kW.

Figure 13.   Rendering of Tactical Laser System (The Cool Gadgets, 2011)

C.   COMPARING ATTRIBUTES OF CURRENT NAVAL LASER SYSTEMS

The attributes of the four developmental laser systems are tabulated in Table 4. Along with the data from Table 1 and Table 4 analysis is done on the four systems.

	SSSL	FEL	FL	
System	Maritime Laser Demonstration (MLD)	Free Electron Laser System (FELS)	Laser Weapon System (LaWS)	Tactical Laser System (TLS) (single phase laser, without laser combination)
System Integrator	Northrop Grumman: Main Contractor NSWC and NAWC: Government test team	Office of Naval Research with several Naval research organization	Directed Energy Warfare Office (DEWO): System Integrator Naval Research Laboratory and Pennsylvania State Electronic-Optic Center: Laser designer Raytheon: CIWS integration effort	Boeing: Beam Director and Laser Weapon Module BAE Systems: MK 38 Machine Gun system integrator
Beam Power	105 kW by coherently combining 7 laser beams of 15 kW each	14.7 kW	33 kW by incoherently combining 6 laser beams	10 kW
Beam Quality	< 3	~ 1	17	2.1
Wall-plug Efficiency	20-25%	10%	25%	30%
Wavelength	1.064 µm	Tunable wavelengths	1.064 µm	Not available
Power Requirement	400-500 kW	10 MW	400 kW	75 kW
Wall Plug Efficiency	20-25%	10%	25%	30%
Technology Readiness Level (TRL)	Approaching 6 (December 2010)	4 (December 2010)	5 (June 2010)	≤ 4 (2011)
Testing Progress	Tested in sea environment	Laboratory testing	Tested in sea environment	Laboratory testing

Table 4.   Comparison of Laser Systems (After O'Rourke, June 2012)

1.   Beam Power

While the beam power of FELS is 14.7 kW, it is assessed that its power can be scaled to megawatt level without there currently being a need for new technical breakthroughs. For both solid-state slab lasers and fiber lasers, the power from each slab or fiber is in the low kilowatt level. However, they have the potential of being scaled up by combining several pieces of slabs or fiber together, as shown in MLD and LaWS respectively.

2.   Beam Quality

A BQ of 1.1 to 5 is considered high for the Navy, whereas a BQ of 5.1 to 20 is deemed moderate (O'Rourke, 2012).

Being near to a diffraction-limited Gaussian beam, i.e., BQ approximately 1, FELS has the best BQ among the three types of lasers. This is one of the many advantages of FEL in general.

Although TLS (FL) has a low beam power as compared to the rest, it has a relatively good BQ of 2.1. This is because it uses only a single phase beam i.e., without beam combining. With incoherent beam combining, the beam quality would likely be affected. It can be observed from the LaWS, which uses incoherent beam combining technology to achieve a higher beam power. Incoherent combining produces output beams with different phases and spanning a significant optical bandwidth. Using this method would reduce the spectral brightness of the beam and affect the BQ.

The MLD (SSSL) coherently combines the laser beams from seven slabs to produce a single beam with the same phase. Unlike incoherent beam combining, this not only produces a higher power output beam, but also ensures that the beam brightness and BQ are not affected significantly.

The longer the range of engagement, the more important BQ becomes, as beam divergence increases over distance. In general, the longer the range to the target, the more important BQ becomes.

3. Wavelength

Only the FELS allows wavelength tuning to match the atmospheric transmission window "sweet spots." With this flexibility, the performance would be less affected by the atmospheric absorption effects. Additionally, wavelength that is less hazardous to human eye (above 1.5 µm) can be selected, although there will be a loss of efficiency with this wavelength.

The wavelength of the MLD, TLS and LaWS are in the region of 1 µm. Development is underway on the three systems to emit light at wavelengths above the more dangerous threshold, i.e., above 1.5µm.

4. Laser Wall-Plug Efficiency

In general, the wall-plug efficiency of laser systems is low. Among the four systems, the FELS has the lowest wall plug efficiency of 10% (Table 4) Consequently, more waste thermal energy would be produced, which would affect laser performance. To minimize overheating, heat has to be dissipated and this would require additional resources (e.g., cooling system) from the naval ship.

5. Weight and Dimension

The estimated weight and dimension of a typical SSSL, FL and FEL are provided in Table 5. The weight and dimension do not consider the power supply subsystem.

	Solid State Slab	Free Electron	Fib
Power	100 kW	100 kW – 1 MW	33 kW
Dimension of laser equipment **below** open deck (Note 1)	1.2 m x 1.3 m x 1.2 m = 2 m^3 (O'Rourke, 2012)	4 m x 4 m x 30 m = 480 m^3 (Sprangle, Ting, Penano, Fischer, & Hafizi, 2008)	1.2 m x 1.3 m x 1.2 m = 2 m^3 (Note 4)
Weight (Assuming 1m^3 weighs ~100 kg)	200 kg	38,400 kg (Anderson, 1996) (Note 3)	200 kg
Dimension of beam control subsystem **on** open deck (Note 2)	1.5 m x 1.5 m x 2.5 m = 5.7 m^3 (Estimated based on the photographs of MLD, reference to the size of the standard 20-foot container. This is assumed to be similar for all three types of laser systems with the same power)		
Weight (Assuming 1m^3 weighs ~100 kg)	600 kg		

Note 1: Equipment under open deck refers to subsystems such as laser equipment racks, which do not need to be in the open.
Note 2: Equipment on open deck refers to beam director and tracking subsystem, etc., that require line of sight to the target, as well as cooling system which needs to be in close proximity to the beam director.
Note 3: The estimated dimension and weight (from different references) of the FEL may not correspond to the same design. However, these values provide the ballpark for the purpose of comparison with the other laser systems.
Note 4: This information for the fiber laser system is not available and is estimated to be similar to the MLD's.

Table 5.   Estimated Dimension and Weight of Laser System

The physical size and weight of the FEL make it extremely challenging or nearly impossible to be installed on any existing Navy ship, without removal of existing equipment and payload. To reduce the size of the FEL, superconducting acceleration structures can be considered. However, this would likely require cryogenic equipment to cool the superconducting structure, which would further increase the already high cost and complexity.

In contrast, the FL and SSSL offer system sizes that are similar to current weapon systems. SeaRAM, which has the same footprint as Phalanx CIWS, weighs about 7,000 kg and has a footprint of 8 m^2 (Raytheon Company, 2006). This makes installation and integration to ship more manageable. Even with additional subsystems needed for beam combining to scale up the power, the subsystems can be stacked (as shown in Figure 14) to optimize space on the ship.

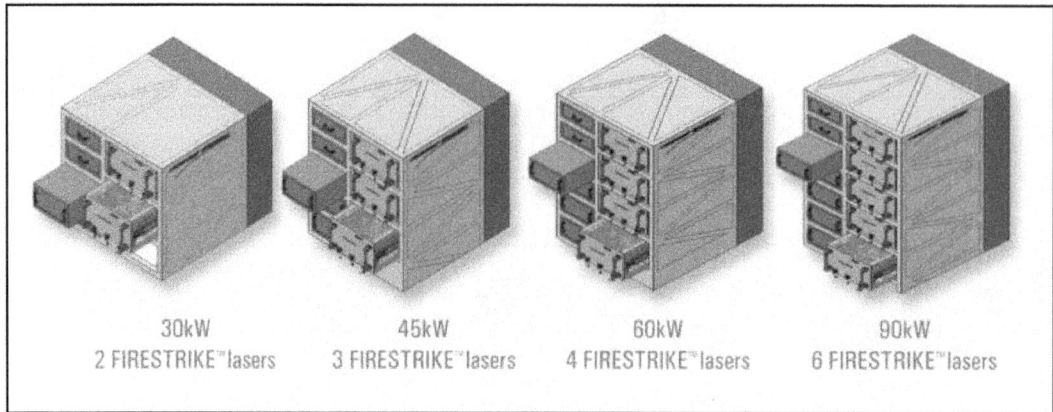

Figure 14.   Scaling Slab Lasers by Stacking (Defense Update , 2012)

It is unlikely that the FEL can be fitted on naval ships in the near term, as novel innovation is required to reduce the size. On the contrary, the up and rising SSSL and FL seem promising, especially with beam combining to enable power scaling.

D.   SINGLE LASER BEAM POWER TREND

Due to the small beam power of an individual SSSL or FL beam, beam combining appears to be paramount to power scaling for the SSSL and FL. However, beam power of a single laser would also affect the degree that beam combining can be achieved.

With the more promising SSSL and FL for near term, Figure 15 offers the laser power progress for both single SSSLs and single FLs from 2000 onwards, as well as the possible trends of these two laser technologies based on the extrapolation of the data points.

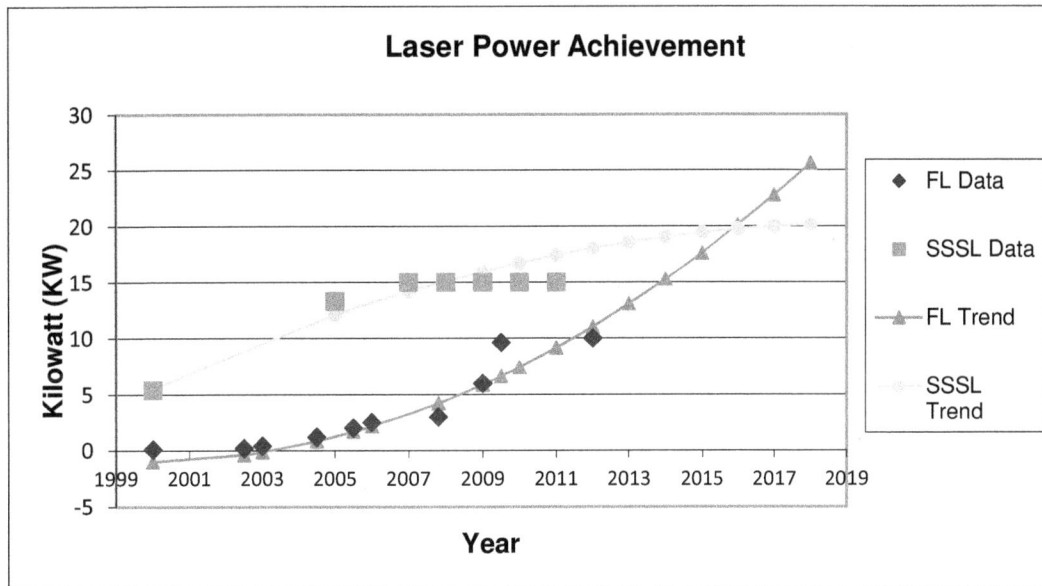

Figure 15.   Trend of SSSL and FL (After Sprangle, Ting, Penano, Fischer, & Hafizi, 2008), (Overton, 2009), (Northrop Grumman, 2008), (Defense Update, 2012)

Clearly, SSSL is ahead of FL in terms of beam power, shown by points on "FL Data" and "SSSL Data." However, from the lines "FL Trend" and "SSSL Trend," it can be observed that the power for the SSSL has seemingly reached a steady state whereas the power for the FL is showing trend of continuous increment, demonstrating the latter's potential in surpassing the SSSL beam power.

The trends presented in Figure 15 are in concert with real-life. Other than military applications, development of a high power FL is also fuelled by commercial applications such as long-distance free-space optical communications and laser-based manufacturing. While for SSSL research, it is primarily driven by defense applications.

The projection shown in Figure 15 is purely a simplified estimation using a second order polynomial which does not take into consideration various intricacies involved. Innovations and breakthroughs are typically spontaneous and discontinuous.

THIS PAGE INTENTIONALLY LEFT BLANK

V.   LASER WEAPON EMPLOYMENT CONCEPT ON A NAVY SHIP

With the research and analysis done on the three types of laser weapon systems in Chapter IV, this chapter specifically assess the feasibility of employing a laser weapon system from a Littoral Combat Ship class of ship against small and fast boat targets. To determine feasibility, the problem statement would be first defined. This entails the operational scope for such weapon. Subsequently, the requirements of a laser weapon system will be identified to meet this operational need. Lastly, capacity of the target platform to carry and operate such weapon system would determine the feasibility.

A.   PROBLEM STATEMENT

As mentioned earlier, Navy forces are facing fast and small targets today. This section of the thesis puts forth the scenario where a LCS or other ship in her ward is being attacked by multiple (up to three targets) explosive-laden suicidal speed boats. This constitutes the dual role of both self-defense and area defense. Some assumptions were made on the target boat shown in Table 6.

Properties	Values
Length	10 meters
Capacity	~400 kg (T.B. Racing and Marine, 2011)
Speed	80 knots (148 km/h = 42 m/s)
Material	Fiberglass / Aluminium
Explosive	Improvised Explosive Device (IED) made of Ammonium Nitrate and Fuel Oil (ANFO) or equivalent Trinitrotoluene (TNT)

Table 6.   Assumptions of the small boat (After Tunaley)

The explosive capacity of the boat is similar to that of a full size car (500 kg) which requires an outdoor evacuation distance of 800 m (U.S. Department of Homeland Security). The explosive effect on sea would be less severe than on land and hence this assumption is reasonable.

As a combat ship, the LCS class ship would have some form of protection against explosion. This means that it could afford to withstand the effects of explosion even if the threat blows up within 800 m of the ship. However, in order for the Navy ship to be safe from any explosion effects, the safety distance is set as 800 m for subsequent calculations.

To disable the small boat, the laser beam can be illuminated on any of these three areas of the boat: the operator, IED and the hull of the boat. While the operator is the most vulnerable among the three, this would not be discussed in this thesis to avoid ramifications against humane treatment.

Figure 16 shows four potential areas of the target boat that can be engaged. The most direct approach would be lasing the beam on the IED, if it is exposed, causing it to explode (Approach A). The next approach is to penetrate through the hull of the boat to reach the IED (Approach B and C). The last is to destroy the motor of the boat (Approach D).

A: Engaging fast boat with exposed IED

B: Engaging fiber glass hull fast boat with hidden IED

C: Engaging aluminum hull fast boat with hidden IED

D: Engaging out-board motor of fast boat with hidden IED

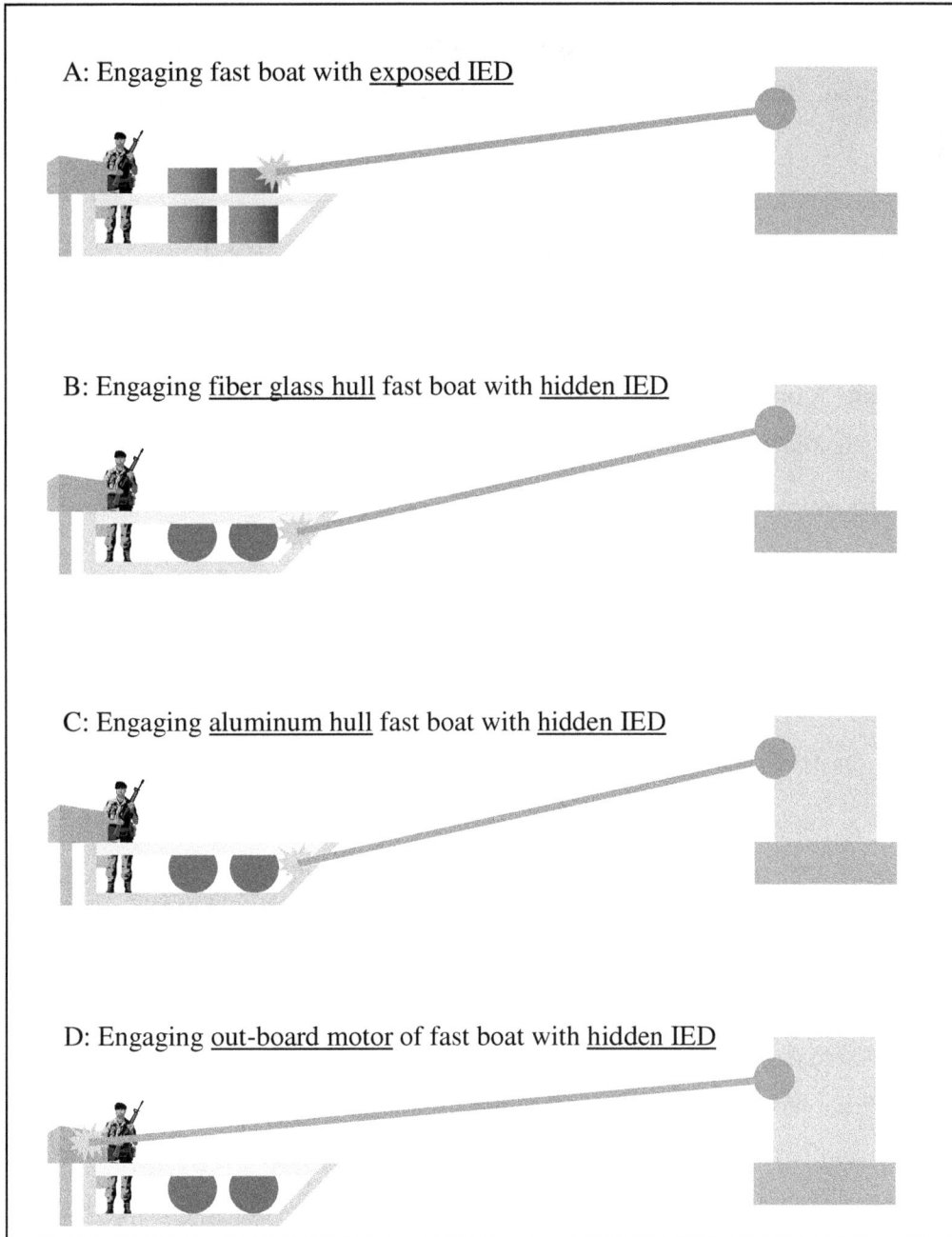

Figure 16.   Areas of a small boat that can be attacked

B. LASER WEAPON REQUIREMENTS

This chapter discusses the requirements for a working laser weapon. These could be generalized into five areas: power, beam size and attenuation, infrastructure, stability and cost.

The laser beam needs to have ample *power* to disable the expected class of targets. To deliver enough energy to the targeted spot, the laser requires sustained lasing. Along with the beam power, the *beam spot size* would also determine the time that the beam takes to disable the target. Due to atmospheric attenuation which increases over range, power losses would result. To account for such power losses, a larger laser power at source is required.

The laser weapon system to be employed on the Navy ship would require various ship *infrastructure capacities* such as space, weight and power supply. Additionally, the system, when installed on the ship, needs to have a certain level of *stability* to accurately sustain continuously lasing on the same spot of the target.

As the laser weapon has to handle multiple (up to three) targets, the system needs to *respond* fast enough to disable all the targets.

An estimation of the required *system acquisition cost* would also be provided. All these requirements would be discussed in detail in the following sections.

1. Power Requirement

To counter the small boat or to damage it physically via hardkill (i.e., to destroy/disable the target such that it could no longer continue its mission), the power of the laser beam has to be strong enough to initiate the on board IED. A common material for making an IED is Ammonium Nitrate and Fuel Oil (ANFO). Ignition of the IED can be achieved by heating the ammonium nitrate present in the IED to its melting point, causing an explosion. The explosion effect is enhanced by the fuel oil present in the IED. With the information from Table 6, the power required to counter the small boat is calculated as follows (Harney, 2012):

To increase the temperature of the material to its melting point, the energy fluence (energy per unit area), F can be estimated as

$$F = \left\{ C_P \left(T_{MELT} - T_{AMBIENT} \right) + \Delta H_{FUSION} \right\} \rho h \qquad \text{Equation 5-1}$$

The description of the symbols and their values for different materials are provided in Table 7. The ambient temperature, $T_{AMBIENT}$, is taken as 300 K.

	Description	Approach A		Approach A/C		Approach B
		Ammonium Nitrate	TNT	Polyethylene	Aluminium (Harney, 2012)	Resin (Note 1)
C_P	Specific heat of material	1.7 J/g-K (Palgrave, 1991)	1.38 J/g-K (Doro-on, 2012)	1.7 J/g-K (Ashby & Johnson, 2012)	0.528 J/g-K	1.6 J/g-K (Ramroth, 2006)
T_{MELT}	Melting temperature	443 K (Gowariker, 2009)	354 K (Harney, 2012)	408 K (Kent, 2003)	1660 K	444 K (Andrew, 1998)
ΔH_{FUSION}	Heat of fusion of the material	76.3 J/g (Army, 1984)	96.6 (Doro-on, 2012)	290 J/g (Scheirs, 2000)	435 J/g	106 J/g (Benedikt, 1999)
ρ	Material density	1.29 g/cm^3 (Gowariker, 2009)	1.64 g/cm^3 (Harney, 2012)	0.95 g/cm^3 (Olabisi, 1997)	4.5 g/cm^3	1.15 g/cm^3 (Mildenberg, Zander, & Collin, 1997)
h	Material thickness	5 mm (Note 2)		5mm (U.S. Plastic Corp.)	1.3 mm (Winger, 2007)	2 mm (Vigor, 2003)
F	Calculated Fluence	206 J/cm^2	140 J/cm^2	225 J/cm^2	675 J/cm^2	78 J/cm^2

Note 1: When resin melts, it would weaken the structure of the fiberglass. Hence in this case, the values of resin are used instead.

Note 2: It is assumed that melting 5mm of ammonium nitrate or TNT would initiate the explosion.

Table 7.   Values of different types of materials

46

While it is easier to attack the weakest point of the boat from the perspective of fluence required, the weakest point may not result in the fastest destruction of the target. For instance, lasing on the fiberglass boat requires the least amount of energy. However, the laser would only melt the resin of the fiberglass, creating a small hole on the hull. It would take a while for the hole to enlarge to a size big enough to sink the boat. During the duration prior to the sinking of the boat, the boat could still crash onto the Navy ship given its fast travelling speed. In this aspect, lasing on the explosives or motor of the boat seems to be a better alternative.

Hitting the explosives would likely lead to an explosion of the boat, preventing the latter from moving closer to the ship. The explosives are assumed to be contained in polyethylene drums for ease of loading onto the boats. Hence prior to the direct lasing on the explosives, the beam has to penetrate through the polyethylene drums. The fluence needed to melt the drum and explosives is similar. For simplicity, a fluence of 250 J/cm^2 would be used.

Destroying the motor would halt the target, as these small boats do not have enough mass momentum to continue to drive towards the Navy ship, despite their high travel speed. The motor housing is assumed to be made of polyethylene, which requires a fluence of approximately 250 J/cm^2 to melt it.

Considering a minimum safe incapacitation distance of 800 m (U.S. Department of Homeland Security), it would be reasonable to assume a maximum lasing time of 1 s to destroy the 10 m small boat (travelling at 42 m/s) detected at 1 km (rounded up from 800 m) away. This works out to a power density of 250 W/cm^2 (power density = fluence ÷ time).

Practically, the target, which has a small radar cross section and is possibly disguised as civilian boat, may be detected only when it is near the ship (short range). However, effectiveness of laser weapon is largely increased in terms of power delivery and precision at short range. Hence, the laser would take less than 1 s to disable the nearer (less than 1 km) target.

In addition to fluence, another parameter to be considered would be the reflectivity of the target. Reflection has two components, specular and diffused for any practical surface. Specular reflection is the mirror-like reflection where light from a single incoming direction is reflected at just one outgoing angle. Diffuse reflection, also known as random reflection, is the reflection of light where an incident ray is reflected at many angles. In general, target surfaces are very rough at laser wavelengths. As a result, the diffuse reflection component normally dominates. In fact, in many cases, there may not be any significant specular component. Table 8 tabulates the approximate diffuse reflectivity of different materials.

Material	Diffuse Reflectivity
Matt Black Paint	4 – 15%
Dirty Olive Drab Paint	4 – 15%
Soil	5 – 15%
Brick	15 – 25%
Vegetation (Glossy Foliage)	30 – 70%
Asphalt	10 – 25%
Concrete	10 – 40%
IR Reflecting Paint	30 – 55%

Table 8.   Approximate Reflectivity of different materials at $\lambda = 1.064$ μm (From NATO)

Assuming reflectivity of 15%, the power density becomes approximately 300 W/cm^2 (= 250 W/cm^2 ÷ 85%). In general, materials lose their strength at temperatures below their melting point and therefore, lose their functionality. Hence, the calculated power density of 300 W/cm^2 is considered an overestimation of the actual value required to counter the small boat.

2. Beam Size Requirement

The laser beam can effect greater destruction if it can be focused on a smaller area. Hence the goal is to focus the beam on the smallest possible spot. However, the beam focus would be affected by a couple of factors. With diffraction, the spot size, S, would be larger.

$$S > \frac{\lambda R}{D}$$

Equation 5-2

where λ is the wavelength, R is the target range, and D is the aperture diameter.

The target is likely to be detected and identified only when it is near the ship. In fact, with less loss from attenuation and less jitter at shorter distance, it would only be easier for the laser to engage the target at shorter range. Nevertheless, it would be better to provide for a longer range so that the laser weapon would not be limited to engagement at such a close range only. The analysis is extended to include a maximum range of up to 10 km. Other than for the purpose of area defense, this range would also enable the laser weapon to counter small boats that are equipped with Hellfire-class missiles (with an effective range of about 8 km). Assuming λ as 1 µm, D as 1 m, and R as 10 km, the spot size would not be smaller than 1 cm.

Beam jitter would also affect the spot size. Assuming a jitter value of 10 µrad, which is considered realistic for mobile systems in low altitudes, a spot size of 10 cm (spot area is 78.5 cm^2) should be the best that could be achieved. With this, the minimum power required to disable the boat target (via its exposed explosives or boat motor) at 10 km is 23.5 kW (= 300 W/cm^2 x 78.5 cm^2).

As the laser beam propagates through the atmosphere, there would be power loss due to atmospheric attenuation (attributed mainly to the absorption by gases). From the exponential Beers-Lambert Law (Weichel, 1990), the attenuation of laser power through the atmosphere is

$$\tau(R) = \frac{P(R)}{P(0)} = e^{-\sigma R}$$

Equation 5-3

49

where $\tau(R)$ = transmittance at range R,

P(R) = laser power at R,

P(0) = laser power at the source, and

σ = attenuation or total extinction coefficient (per unit length)

By combining the Kohschmieder formula with relative extinction coefficient for different aerosol models $\gamma_{Model}(\lambda, RH)$ and visible range, V

$$\sigma_{Model}(\lambda, RH) = \frac{\gamma_{Model}(\lambda, RH)}{V}$$
Equation 5-4

The values of $\gamma_{Model}(\lambda, RH)$ under different Relative Humidity (RH) and environment are provided in Table 9.

1.06 µm	Environment			
RH	Maritime	Rural	Urban	Tropospheric
50 %	2.750	1.639	1.843	1.381
70 %	3.016	1.651	1.823	1.389
80 %	**3.423**	1.690	1.811	1.424
99 %	3.709	1.979	2.152	1.686

1.54 µm	Environment			
RH	Maritime	Rural	Urban	Tropospheric
50 %	2.210	0.943	1.174	0.595
70 %	2.543	0.951	1.154	0.602
80 %	3.536	0.990	1.123	0.638
99 %	3.685	1.264	1.381	0.853

Table 9.   Aerosol Model (Harney, 2012)

By using a RH of 80% in maritime environment, and a V of 10 km, $\sigma(1\mu m, 80\%)$ is 0.3423 km^{-1} (= 3.423 ÷ 10km).

To express attenuation coefficient (km^{-1}) in dB/km,

$$10^{-0.1AV} = e^{-\sigma V}$$
Equation 5-5

where A is the attenuation in dB/km.

The attenuation coefficient of 0.3423 km-1 is expressed as 1.5 dB/km.

Figure 17 provides the attenuation values at sea level with respect to the wavelength. Using a wavelength of 1 μm and assuming a rainfall rate of 1 mm/hr, the atmospheric attenuation is about 1 dB/km from Figure 17. Under the scenario whereby there is no rainfall, an attenuation value of 1.5dB/km will be used. In the case of 1 mm rainfall, an attenuation value of 2.5 dB/km (= 1.5dB/km for aerosol + 1 dB/km for 1mm rainfall) would be used.

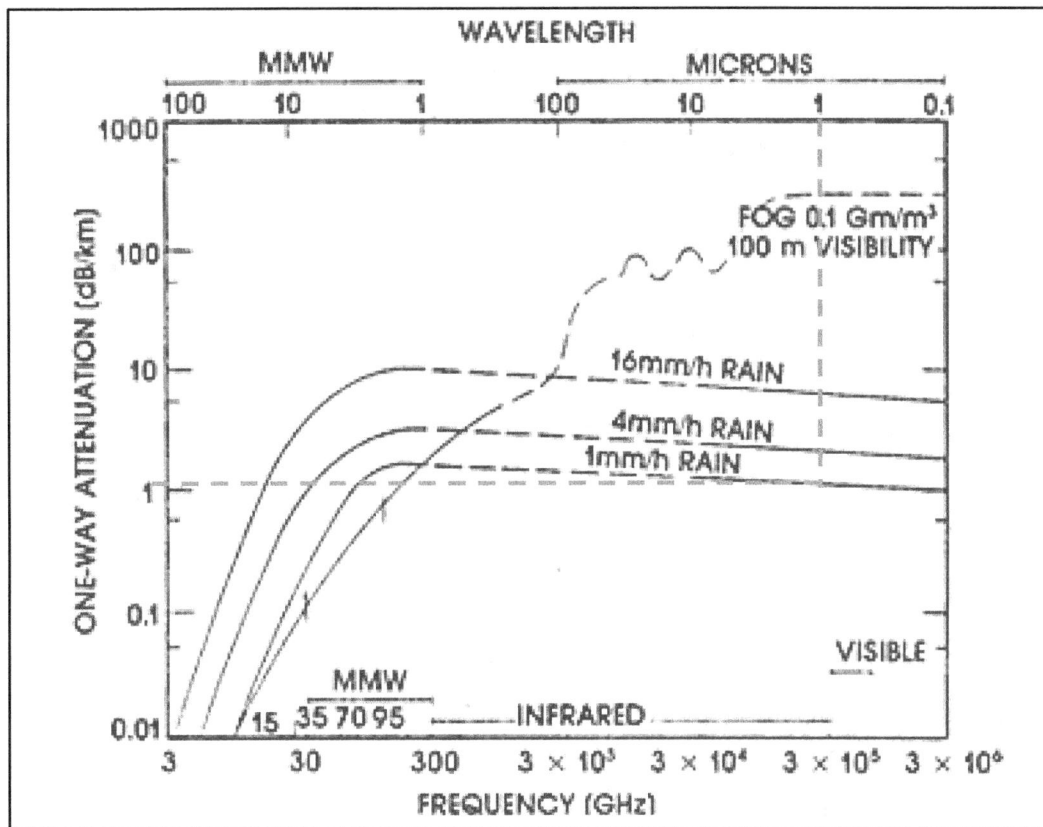

Figure 17.   Atmospheric Absorption at Sea Level (U.S. Department of Transportation Federal Highway Administration)

From earlier calculations, the laser power at R, $P(R)$, needs to be 23.5 kW regardless of the value of R, so as to effect the required destruction on the polyethylene target. Using the above values in Equation 5-3, the required laser power at source at different ranges are tabulated in Table 10.

A further step is taken to calculate the power required to counter a target boat with aluminium hull (Approach C in Figure 16). Assuming an aluminium reflectivity of 85% (Vargel, 2004), the power density becomes 355 kW (= 675 J/cm^2 $\div$ 15% x 1 s x 78.5 cm^2). Using an attenuation of 1.5 dB/km and 2.5 dB/km respectively, the required laser power at source is calculated for different ranges and is also tabulated in Table 10.

Target Range (km)	Without rain		With rain	
	Required Laser Power at Source (kW)			
	Target Material			
	Polyethylene	Aluminium	Polyethylene	Aluminium
1	105	1,591	286	4,325
2	472	7,130	3,488	52,687
3	2,115	31,956	42,489	641,855
4	9,481	143,217	517,622	7,819,395
5	42,489	641,855	6,305,926	95,259,737
6	190,422	2,876,595	76,821,908	1,160,501,167
7	853,414	12,892,003	935,882,433	14,137,798,461
8	3,824,738	57,777,951	11,401,382,092	172,233,644,370
9	17,141,285	258,942,811	138,897,268,481	2,098,235,332,373
10	76,821,908	1,160,501,167	1,692,115,134,429	25,561,739,264,772

Table 10. Required Laser Power at source at different ranges

It can be observed that the required power grows exponentially with range, and the power required to disable the hardened (aluminium) target is much higher than that needed for the polyethylene target. A 1.6 MW of laser source power is required against aluminium material at 1km. Given the same laser source (1.6 MW), it could disable a polyethylene target up to a range of 2 km, in the absence of rainfall. It also is notable that the calculated values in Table 10 are close to those in Table 2 for short range targets (less than 2 km).

For the subsequent sections, the 100 kW, 500 kW and 1.6 MW laser systems would be considered.

3. Ship Infrastructure Requirement

Chapter IV shows that it may not be feasible or practical to install the FEL on Navy ships due to its huge size and high power consumption. Hence in this section, only the SSSL and FL will be discussed.

A laser weapon system can be categorized into three main subsystems: laser equipment, beam control subsystem, and power supply, as depicted in Figure 18.

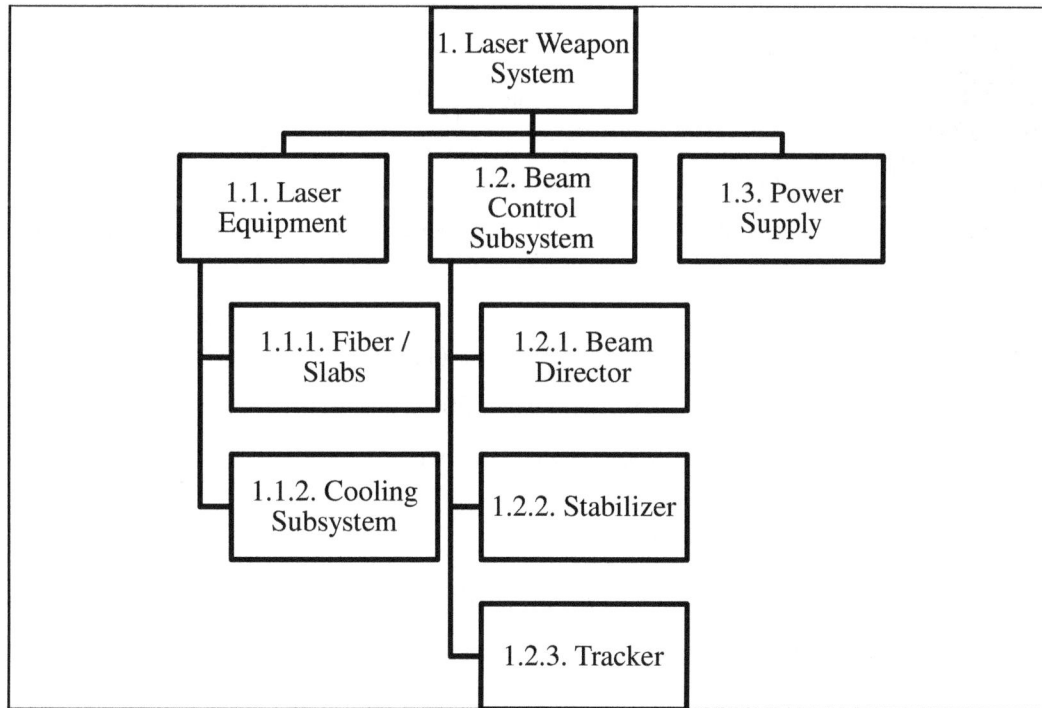

Figure 18.   Subsystems of a typical Laser Weapon System

Due to the lack of details in the system design, a linear relationship is assumed between (a) power and volume and (b) power and weight. The volume and weight of the 100 kW, 500 kW and 1.6 MW laser systems are estimated in Table 11. The values for the 100 kW system are based on the MLD in Table 5. It is also assumed that the weight and dimension of the FL are similar to the MLD's.

	Volume			Weight (Assumes 1m³ weighs 100 kg)		
Subsystem	100 kW	500 kW	1.6 MW	100 kW	500 kW	1.6 MW
On Open Deck						
Beam Control Subsystem	1.5 x 1.5 x 1.5 = 3.4 m³	17 m³	54.5 m³	600 kg	3,000 kg	9,600 kg
Below Open Deck						
Laser Equipment	1.2 x 1.3 x 1.2 = 2 m³	10 m³	32 m³	200 kg	1,000 kg	3,200 kg
Power Supply (reference to Beam Control Subsystem)	6 m³	30 m³	96 m³	600 kg	3,000 kg	9,600 kg
Total (Below Open Deck)	8 m³	40 m³	128 m³	800 kg	4,000 kg	12,800 kg

Note: SeaRAM which has the same footprint as Phalanx CIWS weighs about 7,000 kg and has a footprint of 8 m² (Raytheon Company, 2006).

Table 11.   Estimated Dimension and Weight of SSSL and FL Systems (100 kW, 500 kW and 1.6 MW)

Considering a wall-plug efficiency of 25% (Table 4), a 100 kW system requires an input electrical power of 400 kW, whereas a 500 kW system requires 2 MW. A 1.6 MW system will need 6.4 MW of input electrical power.

4.   Stabilization Requirement

It is pivotal to have a stable beam directed at the target to disable the latter. In a sea environment, the laser system would be subjected to the ship's motion, which would affect the ability of the laser beam to sustain continuous lasing on a single spot. To mitigate this effect, a stabilization subsystem has to be installed.

5.   System Response to Multiple Targets Requirement

In addition to having a stabilized beam, the beam director needs to be able to respond fast enough to counter multiple (up to three) targets that are concurrently approaching the ship at the speed of 80 knots (42 m/s). Based on the earlier assumption that 1 s is required to disable one target, 3 s would be needed to disable three targets. This does not consider the amount of time that the beam director needs to turn from one target to the other.

Figure 19 illustrates the angles at which the next target (Target B1 or B2) could be located with respect to the first target (Target A). The angles are basically classified into two parts, *within* ±90° of Target A (from Target A to Target B1) and *beyond* ±90° of Target A (from Target A to Target B2). Assuming 1 s is required for the beam director to turn from Target A to B1 (*within* ±90°), and 2 s from Target A to B2, (*beyond* ±90°), the time required to disable three boat targets would be in the range of 6 to 9 s. This correlates to a maximum distance of 400 m (= 42 m/s x 9 s) that the targets could travel from the point where the laser system is activated to disabling all three targets.

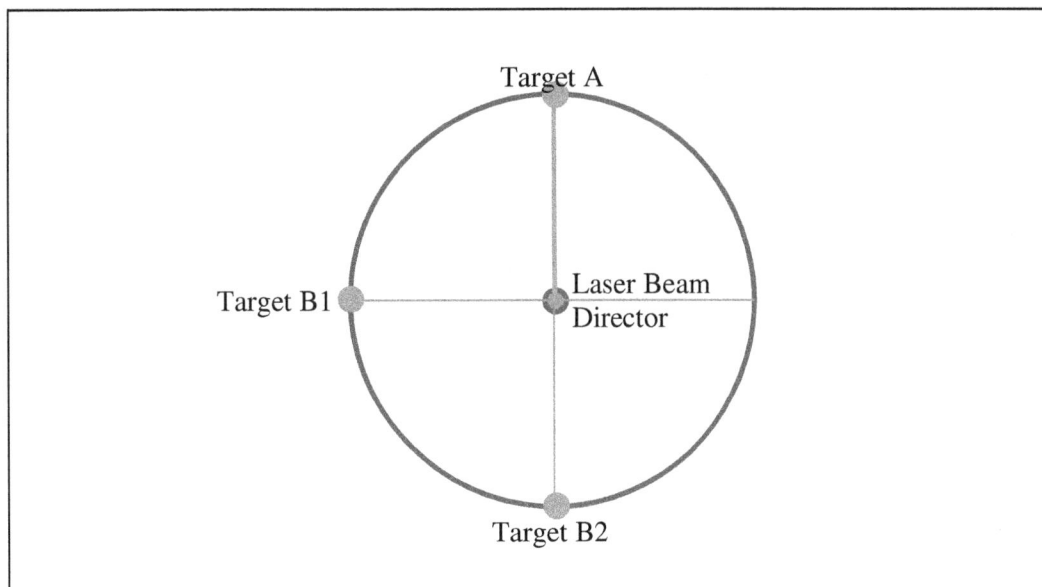

Figure 19.   Location of targets

6.   Cost Requirement

The LaWS (with a potential of scaling up to 100 kW) is estimated to cost approximately $17 million per system (O'Rourke, 2012). However, due to the lack of detailed design information, the actual cost of the 500 kW and 1.6 MW laser systems could not be calculated. Hence for this thesis, cost is assumed to be directly proportional to laser power, which means that a 500 kW laser system would cost $85 million and a 1.6 MW laser system would cost $272 million. As a comparison, a Phalanx CIWS would cost $6M in 2012 dollars (using a 3% cost escalation factor) (GlobalSecurity, 1997).

C. INTEGRATION TO SHIP

Given the requirements of the laser weapon, the feasibility of integrating the system to a Littoral Combat Ship class ship is assessed. Table 12 provides some specification of LCS-1, USS Freedom, built by Lockheed Martin.

Length of Hull	115 m (378 ft)
Displacement	3,089 tons
Speed Sprint	45+ knots (23 m/s)
Electrical Power	3 MW
Mission Bay (Note 1)	594.2 m^2 (6,400 ft^2)

Note 1: The mission bay is one of the key features of the LCS concept, which enables the ship to take on extra equipment tailored to specific missions such as anti-surface or anti-submarine warfare. This extra equipment is packaged into mission modules.

Table 12.   Specifications of LCS-1, USS Freedom (From Jean, 2010)

1. Volume and Weight

The volume and weight of the 100 kW laser system is comparable to that of the Phalanx CIWS's. Hence, installing the system on the ship should be straightforward. Nevertheless, there may be a need to replace the current onboard systems with the laser system, should space availability on the ship pose a constraint.

While the 500 kW and 1.6 MW laser system (Table 11) are both far larger than the Phalanx CIWS, they would be able to fit into the Mission Bay of the LCS (594.2 m^2 from Table 12). Nevertheless, the installation of either system might pose constraints to the ship's deck space (i.e., other systems might have to be removed).

2. Electrical Power

The LCS has four 750 kW diesel generators to provide 3 MW electrical power (Jean, 2010) to all systems onboard the ship.

Referring to the required power at source in Table 10 and using 25% (i.e. four times of the required power at source) wall plug efficiency (Table 4), the power that the laser system needs to draw from LCS to disable the target at different ranges (up to 3 km) is provided in Table 13. This table only reflects the power values for the environment where there is no rainfall.

56

	Polyethylene		Aluminium	
Target Range (km)	At Source (kW)	From LCS (kW)	At Source (kW)	From LCS (kW)
1	**105**	420	**1,591**	6,364
2	**472**	1,888	7,130	28,520
3	2,115	8,460	31,956	127,824

Table 13.   Required ship's Power Supply against Targets in Absence of Rain Condition

A 100 kW laser system requires an electrical input of 400 kW (less than 15% of the ship's supply) in order to disable a target up to 1 km, in the absence of rainfall. This seems acceptable, especially when the laser operation would only occur for a short window.

To counter a target at 2 km in the absence of rainfall, a 500 kW laser system is required. To support this system, 1.9 MW (63% of ship's supply) of ship's power is needed. Although the laser operation will only last for a few seconds, this may not be reasonable, as the ship needs to support other onboard systems (e.g., communication or navigation systems) that are required for the ship's operation.

A 1.6 MW system would draw 6.4 MW of power from the ship. This far exceeds that of the ship's supply (3 MW).

A way to get around the ship's power supply constraint (for the 500kW and 1.6 MW systems) could be using energy storage (e.g., external batteries, capacitors, flywheel, etc.). In this section, lead acid batteries are considered to provide 2 MW (round up from 1.9 MW) and 6.5 MW (round up from 6.4 MW) power supply to the laser system respectively.

In earlier section, it is assumed that 1 s is required to disable a target. With three targets, 3 s is then required. Considering that the laser beam director needs to turn before lasing on the next target, it is assumed that in the worst case, the total lasing time will take 6 s.

Considering the 2 MW power supply, an energy of 12 MJ (= 2 MW x 6 s) is required. Assuming the ship's generator takes 10 min (600 s) to fully charge the lead acid battery, the power drain on the generator would only be 20 kW (=12 MJ ÷ 600 s). With an energy density of around 90 kJ/kg and volumetric energy density of 180 MJ/m^3 (Coley, 2008), a lead acid battery of *140 kg* (= 12MJ ÷ 90 kJ/kg) and *0.1 m^3* (= 12MJ ÷ 180 MJ/m^3) would be adequate to support the 500 kW laser system.

Using the same approach, a 6.5 MW power supply requires an energy of 39 MJ (= 6.5 MW x 6 sec). Given that the ship's generator takes 600 s to charge the battery, the power drain on the generator works out to be 65 kW (= 39 MJ ÷ 600 sec). Such a lead acid battery would weigh *440 kg* (= 39 MJ ÷ 90 kJ/kg) and occupies *0.22 m^3* (=39 MJ ÷ 180 MJ/m^3). This battery would be sufficient to provide power to the 1.6 MW laser system during its short operation.

D.   ANALYSIS

In order for the Navy ship to be safe from any explosion effects, the safety distance is set as 1 km (round up from 800 m).

With the employment of the 100 kW laser system, the current power supply on the LCS-class ship should be able to disable small and fast target at 1 km away, in the absence of rain. While the existing LCS-class ship would be able to meet the requirements (footprint and weight) of the 100 kW laser system, there may be a need to replace current weapon systems onboard with the laser system, should space availability on the ship pose a constraint.

To ensure that the ship could disable targets with aluminium hull at 1 km as well, a 1.6 MW laser system is required. From the weight and dimension perspective, it would be possible to install the 1.6 MW laser system on the LCS-class ship. This may nevertheless compromise other onboard systems which have to be further studied. The power required of the 1.6 MW system exceeds the amount that the ship could realistically provide. As a get around, lead acid battery could be used to store energy required for the short mission. Additionally, with this lead acid battery, it would also be able to support the mission against polyethylene targets at 2 km, in the absence of rainfall.

VI. CONCLUSION AND RECOMMENDATION

Small and fast boat threats laden with explosives are a very relevant threat in the modern era of asymmetrical warfare and terrorism. The ability to employ the power of laser weapons would allow the Navy to be more effective in protecting itself, its friendly forces and innocent civilians.

The scenario of suicide attacks from multiple (up to three) fast motor boats equipped with IEDs against LCS-class ships is used to determine the power required to engage such adversaries. Two sub-scenarios are considered, lasing on the exposed IED that the target boat carries and lasing directly on the target's hull.

The calculations and analysis determine that a 100 kW power laser would be adequate against boats targets carrying exposed IED up to a range of 1 km in an environment without rainfall. The system size would be similar to a Phalanx CIWS and has electrical power supply requirements of about 400 kW. This system would be deemed feasible for employment on an LCS ship.

If the boat target has IED hidden within its aluminum hull, it would require a 1.6 MW power laser at a range of 1 km to burn through the hull to ignite the IED. With a wall-plug efficiency assumed at 25%, the system would require an electrical power requirement of 6.5 MW, which exceeds the ship's electrical power at 3 MW. As a get around, lead acid batteries could be used to store the energy required for this laser system operation. With the weight and volume of the battery taken into consideration, the addition of the lead acid battery to support the operation of the 1.6 MW laser system is assessed to be reasonable. Also with the 1.6 MW laser system, the Navy ship would also be able to counter soft-skin boats targets carrying exposed IED up to a range of 2 km in the absence of rain.

From these studies, it has shown that the current size of the FEL is a major impediment to its practical employment. Until innovation in the FEL reduces overall size and improves robustness, the advantages of its dynamic wavelength tuning and large power output cannot be harnessed on the naval platform (e.g., LCS).

On the contrary, both the FL and SSSL should be able to match the ship's infrastructure capacities (weight, volume, electrical power, etc.). Comparing the SSSL and FL, the SSSL seems to be a more viable choice for near term employment on the LCS platform. This is largely due to its laser power output, which has already reached the 100 kW level.

Although the FL (33 kW) may not yet achieve similar power output levels as the SSSL, it exhibits continual growth. This growth does not solely rely on government funded research, but is also largely pursued by global commercial industries in the fields of telecommunications and laser-based manufacturing.

Therefore, there exists the potential of the FL matching or exceeding the SSSL power levels eventually. With its additional advantages of better cooling and robustness with no free space optics, the FL could better the current SSSL.

LIST OF REFERENCES

Allagaier, G. G. (2003). *The shipboard employment of a free electron laser weapon system* (Master's thesis). Naval Postgraduate School. Retrieved from http://edocs.nps.edu/npspubs/scholarly/theses/2003/Dec/03Dec_Allgaier.pdf

Anderson, E. J. (1996). *Total ship integration of a free electron laser* (Master's thesis) Naval Postgraduate School. Retrieved from http://www.dtic.mil/cgi-bin/GetTRDoc?AD=ADA322386

Andrew, W. (1998). *Polypropylene: The definitive users guide.* Norwich: Plastic Design Library.

Ashby, M., & Johnson, K. (2012). *Materials and design.* Burlington: Elsevier Ltd.

Bass, M. D. (2005). Properties of diode laser pumps for high-power solid-state lasers. *IEEE Journal of Quantum Electronics*, 41, 183-186

Benedikt, G. M. (1999). *Metallocene technology in commercial applications.* Norwich: Plastics Design Library.

Coley, D. (2008). *Energy and climate change: Creating a sustainable future.* England: John Wiley & Sons Ltd.

The Cool Gadgets. (2011, July 27). *Mk 38 Mod 2 tactical laser system: Hybrid of solid state laser weapon and Mk 38 machine gun system.* Retrieved from http://thecoolgadgets.com/mk-38-mod-2-tactical-laser-system-hybrid-of-solid-state-laser-weapon-and-mk-38-machine-gun-system/

Department of Army. (1984). *Military explosives.* Washington, D.C: Headquarters, Department of Army.

Defense Update. (2009). *Firestrike high power solid state laser fires 105kW beam.* Retrieved from http://defense-update.com/newscast/0309/firestrike_laser_190309.html

Defense Update. (2011, August). *Update: Matrix tactical laser weapon demonstrates counter-swarm techniques.* Retrieved from http://defense-update.com/20110830_mk30mod2_tactical_laser_systems.html

Defense Update. (2012, May). *Gamma laser demonstrates burning through an anti-ship missile skin.* Retrieved from http://defense-update.com/20120502_gamma-laser-demonstrates-burning-through-an-anti-ship-missile-skin.html

Defense Update. (2012). *Northrop Grumman introduces a new weaponized solid-state laser.* Retrieved from http://defense-update.com/20081115_firestrike_151108_laser.html

Department of Homeland Security. (n.d.). *IED Attack.* Retrieved from http://www.dhs.gov/xlibrary/assets/prep_ied_fact_sheet.pdf

Directed Energy Weapons. (2007). Retrieved from Defense Science Board Task Force website: http://www.acq.osd.mil/dsb/reports/ADA476320.pdf

Doro-on, A. (2011). *Risk assessment for water infrastructure safety and security.* Boca Raton: CRC Press.

Fan, T. Y. (2005). Laser beam combining for high power, high-radiance sources. *IEEE Journal of Selected Topics in Auantum Electronics*, 11, 567-577.

FAS Space Policy Project. (1998, March). *Mid-infrared advanced chemical laser.* Retrieved from http://www.fas.org/spp/military/program/asat/miracl.htm

GlobalSecurity. (1997). *Phalanx close-in weapon system.* Retrieved from http://www.globalsecurity.org/military/library/budget/fy1997/dot-e/navy/97ciws.html

Gowariker, V., Krishnamurthy, V. N., Gowariker, S., Dhanorkar, M., Paranjape, K., & Borlaug, N. (2009). *The fertilizer encyclopedia.* Hoboken: John Wiley and Sons, Inc.

Harney, R. C. (2012). *Lasers systems.* Monterey: Naval Postgraduate School

Hecht, J. (2007). *Photonic frontiers: Laser weapons - pumping up the power.* Retrieved from Laser Focus World website: http://www.laserfocusworld.com/articles/print/volume-43/issue-5/features/photonic-frontiers-laser-weapons-pumping-up-the-power.html

Horn, A. (2009). *Ultra-fast material metrology.* Weinheim: Wiley-VCH.

Hunter, B. V., & Leong, H. K. (1996). *Understanding high-power fiber-optic laser beam delivery.* Argonne: Argonne National Laboratory. Retrieved from http://www.ne.anl.gov/facilities/lal/Publications/BeamDelivery/JLABeamDeliveryManuscrip.pdf

Jean, G. V. (2010, March). *Builders of the navy's littoral combat ship pull out all the stops.* Retrieved from National Defense Industrial Association Technology Magazine website: http://www.nationaldefensemagazine.org/archive/2010/March/Pages/LittoralCombatShip.aspx

Jefferson Lab News. (2006, April). *Free-electron laser targets fat.* Retrieved from
http://www.jlab.org/news/releases/2006/fel.html

Kent, J. A. (2003). *Riegel's handbook of industrial chemistry.* New York: Plenum
Publishers.

McDermott, W. E., Pchelkin, N. R., Bernard, D. J., & Bousek, R. R. (1978). An
electronic transition chemical laser. *Applied Physics Letters*, 32, 469-470.

Mellish, B. (2005). *Population inversion 3-level diagram.* Retrieved from Wikipedia
website: http://en.wikipedia.org/wiki/File:Population-inversion-3level.png

Mildenberg, R., Zander, M., & Collin, G. (1997). *Hydrocarbon resins.* New York: VCH.

Motes, R. A., & Berdine, R. W. (2009). *Introduction to high-power fiber lasers.*
Albuquerque: Directed Energy Professional Society.

National Research Council, C. o. (1994). *Free electron lasers and other advanced
sources of light: scientific research opportunities.* Washington, D.C.: National
Academy Press.

NATO, O. (n.d.). *Laser systems performance.* Retrieved from
http://ftp.rta.nato.int/public//PubFullText/RTO/AG/RTO-AG-300-V26///AG-300-
V26-03.pdf

NITEHOG Systems LLC. (2012). *How thermal works.* Retrieved from
http://www.nitehog.com/how-thermal-works.html

Northrop Grumman. (n.d.). *Joint high power solid-state laser (JHPSSL) program.*
Retrieved from http://www.as.northropgrumman.com/
products/joint_hi_power/index.html

Northrop Grumman. (2008). *Solid-state laser weapons.* Retrieved from
http://www.as.northropgrumman.com/products/ssl/assets/SSL_Datasheet_111008.
pdf

Office of Naval Research. (2008). *100 kW FEL broad agency announcement.* Retrieved
from http://www.onr.navy.mil/~/media/Files/Funding-Announcements/BAA/08-
013.ashx

Olabisi, O. (1997). *Handbook of thermoplastics.* New York: Marcel Dekker, Inc.

O'Rourke, R. (June 2012). *Navy shipboard lasers for surface, air, and missile defense:
Background and issues for Congress.* Retrieved from Congressional Research
Service website: http://www.fas.org/sgp/crs/weapons/R41526.pdf

Overton, G. (2009, June). *IPG Photonics offers world's first 10 kW single-mode production laser*. Retrieved from Laser Focus World website: http://www.laserfocusworld.com/articles/2009/06/ipg-photonics-offers-worlds-first-10-kw-single-mode-production-laser.html

Palgrave, D. A. (1991). *Fluid fertilizer science and technology*. New York: Marcel Dekker. Inc.

Paschotta, R. (2008). *Encyclopedia of laser physics and technology*. Berlin: Wiley-VCH.

Ramroth, W. T. (2006). *UMI microform*. Ann Arbor: Proquest Information and Learning Company.

Raytheon Company. (2006). *SeaRAM evolved ship defense*. Retrieved from http://www.raytheon.com/capabilities/products/stellent/groups/public/documents/content/cms01_055726.pdf

Rutherford, T. S. (2001). Yb:YAG and Nd:YAG edge-pumped slab lasers. *Optics Letters*, 26, 986-988.

Scheirs, J. (2000). *Compositional and failure analysis of polymers*. England: John Wiley and Sons, Ltd.

Sprangle, P., Ting, A., Penano, J., Fischer, R., & Hafizi, B. (2008). *High-power fiber lasers for directed-energy applications*. Retrieved from Naval Research Laboratory website: http://www.nrl.navy.mil/content_images/08FA3.pdf

Stephen, M. A. (2008). *Fiber-based laser transmitter and laser spectroscopy of the oxygen A-band for remote detection of atmospheric pressure*. Ann Arbor: ProQuest LLC.

Svelto, O. (1979). *Principles of lasers*. Milan: Plenum Press.

T.B. Racing and Marine. (2011). *New Boats*. Retrieved from http://www.tbracingandmarine.com/shop/index.php/new-boats.html

Tunaley, J. K. (n.d.). *Smuggler and pirate go-fast boats*. Retrieved from London Research and Development website: http://www.london-research-and-development.com/GOFAST.pdf

Tyler, A., Korczynski, E., & Sumantri, K. (2000, March 1*). Diode-pumped solid-state lasers: Edge pumping drives slab-laser performance*. Retrieved from Laser Focus World website: http://www.laserfocusworld.com/articles/print/volume-36/issue-3/features/diode-pumped-solid-state-lasers-edge-pumping-drives-slab-laser-performance.html

UCLA Laser Safety Lite. (2009). *Laser lite: A quick overview of laser safety.* Retrieved from http://ehs.ucla.edu/Pub/RSD/LSM%20Lite.pdf

U.S. Department of Homeland Security. (n.d.). *IED attack, improvised explosive devices.* Retrieved from http://www.dhs.gov/xlibrary/assets/prep_ied_fact_sheet.pdf

U.S. Department of Transportation Federal Highway Administration. (n.d.). *Weather applications and products enabled through vehicle infrastructure integration.* Retrieved from http://ops.fhwa.dot.gov/publications/viirpt/sec5.htm

U.S. Plastic Corp. (n.d.). *Heavy duty plastic open head drums.* Retrieved from http://www.usplastic.com/catalog/item.aspx?itemid=22740&catid=459

Vargel, C. (2004). *Corrosion of aluminium.* Oxford: Elsevier Ltd.

Vigor, J. (2003). *The practical encyclopedia of boating.* Blacklick: McGraw Hills Companies.

Weichel, H. (1990). *Laser beam propagation in the atmosphere.* Bellingham: SPIE.

Williams, B. W. (2005). *Jefferson Lab Free Electron Laser 10kW upgrade – Lessons learned.* (Master's thesis) Naval Postgraduate School. Retrieved from http://www.dtic.mil/dtic/tr/fulltext/u2/a434107.pdf

Willingale, R. (2007). *Lasers and quantum optics.* Retrieved from University of Leicester, Department of Physics and Astronomy website: http://www.star.le.ac.uk/~rw/courses/lect4313.html

Winger, J. (2007). *Small boats mechanics and materials.* Bloomington: AuthorHouse.

THIS PAGE INTENTIONALLY LEFT BLANK

INITIAL DISTRIBUTION LIST

1. Defense Technical Information Center
   Ft. Belvoir, Virginia

2. Dudley Knox Library
   Naval Postgraduate School
   Monterey, California

3. Professor Robert C. Harney
   Naval Postgraduate School
   Monterey, California

4. Dr. Douglas Nelson
   Naval Postgraduate School
   Monterey, California